William Samuel Johnson, Winthrop Saltonstall

The Chemical and Medical History of Septon, Azote or Nitrogene

And Its Combinations with the Matter of Heat and the Principle of Acidity

William Samuel Johnson, Winthrop Saltonstall

The Chemical and Medical History of Septon, Azote or Nitrogene
And Its Combinations with the Matter of Heat and the Principle of Acidity

ISBN/EAN: 9783743421967

Manufactured in Europe, USA, Canada, Australia, Japa

Cover: Foto ©berggeist007 / pixelio.de

Manufactured and distributed by brebook publishing software
(www.brebook.com)

William Samuel Johnson, Winthrop Saltonstall

The Chemical and Medical History of Septon, Azote or Nitrogene

THE CHEMICAL AND MEDICAL HISTORY OF SEPTON, AZOTE, OR NITROGENE; AND ITS COMBINATIONS WITH THE MATTER OF HEAT AND THE PRINCIPLE OF ACIDITY.

SUBMITTED TO THE PUBLIC EXAMINATION

OF THE

FACULTY OF PHYSIC;

UNDER THE AUTHORITY OF THE

TRUSTEES OF COLUMBIA COLLEGE

IN THE

STATE OF NEW-YORK:

WILLIAM SAMUEL JOHNSON, LL.D. Prefident:

FOR THE DEGREE OF

DOCTOR OF PHYSIC.

ON THE THIRD DAY OF MAY, 1796.

By WINTHROP SALTONSTALL,

Citizen of the State of Connecticut.

"If the phenomena of animal bodies, as is highly probable, depend upon the matter of which they are compofed, and if their refpective functions be the refult of a peculiar combination of matter; then muft CHEMISTRY chiefly pave the way for A RATIONAL PHYSIOLOGY OF ANIMAL BODIES, becaufe it acquaints us with matter and its conftituent parts."

REIL.

NEW-YORK:

Printed by T. and J. SWORDS, Printers to the Faculty of Phyfic of Columbia College, No. 99 Pearl-ftreet.

—1796.—

PREFACE.

PREVIOUS to commencing the chemical history of the subject of these pages, I must apologize for the assumption of the new term by which I have designated it: this is the more necessary, as a diversity of terms, expressive of the same idea, may be said to retard the advancement of knowledge; and especially as the innovation of an unexperienced student may have the appearance of boldness or presumption: yet the reasonableness and propriety of the thing itself, it is hoped, will amply warrant the adoption. It is well known, that the combinations afforded by *nitre* have been hitherto considered as of mineral origin, and that the terms azote, azotic, nitrous, nitric, &c.* were adopted by the French Academicians; and as the first of these and its derivatives were exceptionable; the term nitrogene had been substituted for it by some of the later writers:† but, from the evidence of the most striking facts, we shall, in the sequel, be able to prove, that the nitrous acid, forming a part of the salt nitre, is of *animal derivation*, formed from azote and oxygene, during the putrefaction of substances of that kind; and that by the addition of the vegetable

* See Table of Chemical Nomenclature, by Lavoisier, &c. for 1787, title Azote.

† Chaptal's Chemistry, passim—Pearson's Translation of the Nomenclature.

vegetable alkali, and that only, is nitre formed. Hence, then, we assume the word septon, to express the radical of this compound, and derive it from σηπω, *putrifacio*, whence comes SEPTON, *putridum*. Thus the combinations formed from this base of the nitrous acid may have new terms in the nomenclature, and be arranged in the following order:—

Septon, for azote or nitrogene.
Septous gas, for azotic gas, or atmospherical mephitis.
Gaseous oxyd of septon, for dephlogisticated nitrous air.
Septic gas, for nitrous gas.
Septous acid, for nitrous acid.
Septic acid, for nitric acid.
Septate, septite, for nitrate, nitrite, &c.

I think it proper here to premise the liberal use I have made of the published sentiments of my friend and instructor, Professor MITCHILL; the attendance on whose valuable lectures, in the College of New-York, has afforded me a large share of materials for the present dissertation; and who first suggested this alteration of the nomenclature.

CONTENTS.

CHAPTER I.

The Chemical History of Septon, and its Combinations, including,

CHAPTER II.

Its Physiological and Medical Operation—comprehending,

Sect.

APPENDIX.

AN

AN

INAUGURAL DISSERTATION

ON

THE CHEMICAL AND MEDICAL HISTORY

OF

SEPTON, AZOTE, OR NITROGENE.

CHAPTER I.

The Chemical Hiſtory of Septon, and its Combinations.

SECTION I.—*Septon, and Septous Gas.*

SEPTON is one of the most abundant elements in nature: combined with caloric, it forms septous (azotic) gas, which composes nearly two thirds of our atmosphere. No experiments hitherto instituted, have detected it in a distinct and separate form; for its attraction for the matter of heat is so great, that in all temperatures of our earth it exists in the gaseous state. It is likewise one of the constituent elements of animal bodies, in combination with carbone, hydrogene, and phosphorus: these are united together by a certain proportion of oxygene, forming animal oxydes and acids, according to the degree of oxygenation. Oxygenous gas forms nearly the other third of our atmosphere: occasionally, however, other gaseous ingredients enter into the composition; but these are to be considered as *accidental*, not *essential* to its constitution; and an atmosphere in which there are other ingredients than oxygenous and septous airs, may always be supposed to obtain, to a considerable degree, in cities and other crouded places, and in marshy and swampy situations.—Septous gas may be procured from the atmosphere, by exposure of a given quantity, to be acted upon by the sulphure of potash (hepar sulphuris), in a close vessel, by which operation its oxygene is absorbed, and the septous

gas

gas left behind by itself. It is afforded by the decomposition of the animal muscular fibre, both by an *artificial* and *natural* process: in the former case by subjecting the muscular fibre to the action of the septic (nitric) acid: during the process, the septous gas is evolved; while the carbone and hydrogene of the muscle run together, forming the oily pellicle floating on the surface. That this process takes place in the natural and spontaneous decomposition of the muscular parts of animals, under certain favourable circumstances, is evinced by the testimony of Mr. Sneyd,* who gives an account of the conversion of a bird into a hard fatty substance. It should seem to have undergone this change by long submersion in the mud and sediment of a fish-pond. An analogous instance, of the escape of this gas from its muscular union, is inferred from the relation of a disorganizing process by Mr. Gibbs, of Oxford,† in his paper on the conversion of animal muscle into a substance much resembling spermacæti; and as also appears from Fourcroi's report on the removing the Cimiterié des Innocens at Paris. In all these cases the original structure of the parts appeared to be the same, though their nature and properties were very much changed by the process they had undergone.

This gas, in a pure form, is incapable, like oxygene air, of sustaining animal respiration ‡ and combustion: at the same time that it and its compounds appear to serve as a *pabulum vitæ* of vegetables, whose œconomy is such as probably to decompound it, and attach septon to themselves, as a nutritive ingredient.

Septous gas, combined with hydrogene gas, forms ammoniac (volatile alkali); and it seems to be the opinion of the ablest chemists, that the other alkalies are like combinations.§

SECT. II.—*Gaseous Oxyd of Septon.*

SEPTON is capable of combining with the principle of acidity, and the compounds, according to the degree of the oxygenating process, are the following: 1. Gaseous oxyd of septon. 2. Septic gas. 3. Septous acid. 4. Septic acid. All which are of a nature so interesting as to merit a particular enumeration and discussion.

This combination of the principle of acidity, with septon and caloric, in the first or lowest degrees, (for an acquaintance with which we are indebted to the indefatigable labours of Priestley, by whom

* Philosophical Transactions, for 1792, part ii.

† Repertory of Arts and Manufactures, No. viii. p. 105.

‡ Priestley's Experiments and Observations on different kinds of Air, vol. ii. p. 205.

§ Chaptal, vol. i. p. 108.

whom it was called dephlogisticated nitrous air) is possessed of some surprising properties.

Previous to its discovery, from the analogy subsisting between *respiration* and *combustion*, it was considered, that what would facilitate the one, would have a like influence upon the other, and *vice versa*, the same causes would impede both. But Doctor Priestley informs us,* that this gaseous substance keeps up combustion naturally and freely in a candle immersed in it; but is, at the same time, highly noxious to animals, and destroys their life the moment they are put into it.† "This surprising quality is doubtless owing to the difference in the attractive force which its oxygene exerts for hydrogene in the one case (combustion,) and for carbone in the other (respiration); for it is known, that by mixing the gaseous oxyd of nitrogene with carbonated hydrogene gas, the carbone is precipitated from its solution. Hence it appears, that the attraction for charcoal is much weaker than for hydrogene, and that although carbone may be made to burn in the gaseous oxyd, hydrogene is the substance for which it has the closest affinity. And we can now readily conceive how the hydrogene of the candle may, in an especial manner, contribute, by attracting the principle of acidity from the gaseous oxyd, to keep up the inflammation, wherein some part of the charcoal may likewise, though in a secondary way, be converted to carbonic gas. It may be understood too, wherefore it is not capable of sustaining life. There are two important purposes answered by animal respiration; the one to furnish oxygene to the phosphoric, sulphureous, and carbonic matter of the blood; the other to carry off its surplusage of charcoal by means of the lungs. Now, the gaseous oxyd has less action upon phosphorus and sulphur, than it has upon charcoal. Hence it is a very natural conclusion, that, in ordinary breathing, the gaseous oxyd does not only not yield its principle of acidity to the blood in the pulmonic circulation, but at the same time does not sufficiently attract carbone from the venous portion of it: whence it comes to pass, that an animal inhaling an air, contributing to neither of these salubrious processes, must speedily die; its blood being both in a disoxygenated and super-carbonated state; hydro-

B gene

* Experiments and Observations on different kinds of Air, vol. ii. p. 55.

† Priestley seems not to have understood the composition of this air, and for a considerable time it attracted but little notice from philosophers: yet, very recently, Messrs. Deimann, Trooſtwick, &c. have prosecuted this subject with much experimental accuracy, have confirmed the experiments of Priestley, related a variety of processes for obtaining it, and declared it to consist of 37 parts of oxygene, chemically united to 63 parts of septon.—Recherches Physico-Chemiques, &c. Amsterdam, 1794.

gene alone being the ingredient, in phlogistic operations, which readily attracts its oxygene from the gaseous oxyd."*

During the plague at Marseilles, (which I shall hereafter endeavour to show was occasioned by the same vitiated state of the atmosphere that causes all pestilential and most epidemic diseases, to wit, the chemical union of septon with oxygene) a great quantity of wood, brush, and faggots were procured, laid in piles at small distances from each other, along the walls of the town, in all the public walks, squares and markets, and at the house doors—set fire to at the same time, and burned to ashes. This, though repeated till all the fuel was spent, did no good. Afterwards, quantities of brimstone were bought up and used in fumigations: but the sickness raged with greater virulence than before.—The experiment made with *tar*, in the city of New-York, in September, 1795, was of no service, and the epidemic increased afterwards. The reason of which is obvious; the gaseous oxyd resists the attractive powers of charcoal and sulphur, and cannot be resolved by them into its constituent parts. Fires made with oil, fat, wax, and tallow, might perhaps have been advantageous, on account of the greater proportion of hydrogene in these substances. During the plague in London, in 1665, large fires of coal were continued, in many parts of the city, for three days, and an unusual mortality immediately ensued. Was this, it might be asked, by consuming the oxygene gas, while the oxyd itself remained unaltered by the coal fires? If so, they are neither useful nor harmless, but of the most pernicious consequence.

This gaseous oxyd can be separated from septous gas by its ready miscibility with water, with which it combines, in its natural temperature, in a very pure form, at the same time that it undergoes no shrinking decomposition, or change, by mixture with the atmospheric fluid, nitrous gas, or vital air: but it is to be remembered, that, by mixing with water, it is only taken out of circulation in the atmosphere, and not decompounded; and on evaporation, or the application of a sufficient degree of heat, it is dissevered from its connection, and re-exists in the atmosphere.

This gaseous oxyd may be obtained by exposure of septic (nitrous) gas to the sulphure of potash (hepar sulphuris): after a certain progress, in the combination of the oxygene and sulphure, has been made, the residuum is found to have no acidity—to be a proper oxyd, and to possess the permanently elastic properties of a gas.

SECT

* Mitchill on Contagion, p. 7 & seq.

Sect. III.—*Septic Gas.*

THE oxygenous principle, uniting with septon in a larger proportion, forms septic (nitrous) gas: this being found to consist of not less than 68 parts of oxygene, *chemically united* to 32 parts of septon, and not unfrequently the acidifying ingredient being yet several degrees more abundant.

From the avidity with which this fluid unites with oxygene, we never shall be enabled to procure it from atmospheric air, but at the instant of its disengagement from its original sources of derivation; for, on being let loose into the atmosphere, it instantly unites with oxygene, after the manner it does in the Eudiometer, and forms the septous (nitrous) acid. It differs further from oxygenous air, as it is incapacitated to support flame; from hydrogene gas, in being incombustible; from carbonic acid gas, in its smaller specific gravity, its specific gravity being even less than atmospheric air,* whereas carbonic acid gas is greater.

"Plants die very soon, both in nitrous air, and also in common air saturated with nitrous air, but especially the former. This kind of air is as noxious as any whatever, a mouse dying the moment it is put into it: but frogs and snails (and probably other animals whose respiration is not frequent) will bear being exposed to it, though they die at length. A frog, put into nitrous air, struggled much for two or three minutes, and moved now and then for a quarter of an hour; after which it was taken out, but did not recover. There is something remarkable in the effect of nitrous air on insects that are put into it. *Wasps* always died the moment they were put into the nitrous air. I could never observe that they made the least motion in it; nor could they be recovered to life afterwards. This was also the case with *spiders*, *flies*, and *butterflies*: sometimes, however, spiders would recover after being exposed about a minute to this kind of air."†

This gas may be obtained by submitting steel filings to the action of the septous (nitrous) acid, by which operation *part* of the oxygene of the acid, uniting with the iron, forms a septate or nitrate of iron, and *the other portion* is evolved in the gaseous form. It is this gas which, from its strong attraction for oxygene, is made use of to ascertain the existence and proportion of the later gas in chemical experiments: but from its great mutability, and diversity of comparative strength, it is a precarious test; for, says the Abbé Fontana, "the nitrous air which is made at one and the same time, in the same flask, with the same nitrous acid, with the same [illegible]

a[illegible]

* Chaptal, vol. i. p. 258.

† Priestley's Experiments and Observations on different kinds of Air, [illegible] p. 409.

and, what is more, the air which comes from the same flask, varies extremely in its strength, and is more or less capable of diminishing common air, and dephlogisticated (oxygene) air; or, to speak more properly, being diminished by these airs, according to the circumstances under which it is collected."*

Sect. IV.—*Septous and Septic Acids.*

THE next degree of combination, in which the acidifying principle exists united with septon, is that of the septous (nitrous) acid. Hitherto the origin of this acid has been deduced from the *mineral kingdom*, especially from the salt called nitre, or the septate of potash. How far this deduction is founded in truth, and how far it is warrantable, may appear from the following facts cited, and inferences drawn, by the Professor of Chemistry, in his late publication on the oxyd of septon. Speaking of nitre, it is remarked, "that substance is known to consist of nitrous acid, joined to potash. It is usually formed during the decay of animal and vegetable bodies, and, by a spontaneous process, is produced from their ruins. We are quite satisfied, that azote (septon) and oxygene entered into the composition of those bodies when alive, and have gone into new combinations on their disengagement by death. One of these recent compounds must be nitrous acid, constituting, by junction with a saline base, the nitrate (septate) of potash. Thus the theory of the formation of salt-petre necessarily presumes the generation of nitrous acid from two of the elements disengaged from organic texture. And as azote, the radical of the acid, is especially abundant in animal bodies, and, as Lavoisier† says, *favorise marveilleusement la putrifaction*, wonderfully promotes putrefaction, there is little difficulty in conceiving, both how, in such circumstances, it attracts the acidifying principle, and afterwards attaches itself to the alkali. But further than this, the authority of Mr. Becker‡ has been advanced in favour of the production of nitrous acid without the aid of the putrifactive fermentation at all. He found nitrous acid in the urine of cows, which had been eight days exposed to the sun. He mixed some of the soakings of a dung-hill with a ley of burnt sheep's dung and chalk in powder. The mixture began to ferment on the following day, and on the fourth, the internal commotion having ceased, he found, at the bottom of the phial, regular chrystals of prismatic nitre. He ascribes the nitrous acid, not to a process going on in the air, but brought

* Recherches Physiques, p. 9.

† 1 Traite Elementaire de Chimie, 155.

‡ Notes to Bergman's Elective Attractions, p. 327.

brought about by the *exertions of animals*. On examining the earth of stables and cow-houses, he found its lixivium to yield prismatic nitre, while that of the dung would afford only small chrystals, which required an addition of nitre in order to be reduced to a prismatic form: and he declares he can attract salt-petre at pleasure, in the course of three days, from the earth of stables and cow-houses, by using, for saturation, well purified pot-ashes. In the production of salt-petre, the putrified substance, if of the animal kind, affords little more than the nitrous acid. This was known to Boerhaave, who says, the nitrous quality of the earth is derived from the excrements of animals and their putrified carcases, particularly such as do not use sea-salt, as birds, which, by the addition of the ashes procured from the burning of plants and of quick-lime, forms salt-petre.* This fact, of the animal origin of the nitrous acid, is confirmed by the testimony of Macquer,† who declares, that in the putrifactive process which affords nitrous acid, animal substances have a decided preference; so that, in order to make a chrystalizable salt-petre from substances purely animal, a quantity of the vegetable alkali must be added; whilst the salt-petre produced in the putrefaction of vegetables alone is naturally found to be furnished with that quantity of fixed alkali, which is necessary to form good nitre." To this may be added the authority of Fourcroi.‡ When, to all this, it is subjoined, that on analizing the soil taken from the bottoms of graves, where human bodies have putrified, it has been found, though having no communication with the external air, to be highly charged with nitrous acid, the animal origin of this acid is put entirely out of doubt. Again,

In a treatise on salt-petre, by James Massey, Esq; § from the learned and perspicuous manner in which the subject is treated, we are enabled to deduce the most ample proofs of the animal origin of this acid, independent of other sources of information. "If we throw a fixed salt into any putrid liquor, it will be nutralized by it: and now, if we dip a piece of soft paper into this mixture, and dry it, it will burn like a match, in the same manner as if dipped into a weak solution of salt-petre; which shows, that it not only contains an acid, but one of the *nitrous* sort."—"But the strongest proofs of the existence of an acid in putrid juices, if the earths of stables and cow-houses do not afford an equal one, must be drawn from the soil at the bottoms of graves, which can certainly

* 1 Elementa Chemie, 44.

† 3 Dictionare de Chemie, 18.

‡ See 2 Lecon. Elementaires, &c. 842.

§ Repertory of Arts and Manufactures, vol. i. p. 248 & seq.

tainly derive its nitrous quality from nothing but the corrupt bodies with which it lies in contact: and this may satisfy us in respect to the source from which other absorbent earths may derive it."— "The common soil, in some parts of India, is naturally nitrous, owing plainly to the fish and slime that are left upon it by the inundations of the river Ganges, which soon corrupt in that hot climate, and fill the earth with putrid juices; and here putrefaction, being carried on with the greatest rapidity, is of course soon completed, and the natives are, in a short time, furnished with a nitrous earth perfectly maturated. But it must not be forgotten, that their strongest earths are found at the bottoms of their tanks, or shallow ponds of water, which, in that country, are often of great extent, and in which, the water being evaporated by the heat of the sun, large quantities of fish are left to corrupt, which furnish a mud of the strongest nitrous quality. In this manner nitrous earths are naturally formed in those parts of the world."

A sensible English traveller * has given an account of the saltpetre works in Grenada, and agrees with this idea of the formation of nitrous acid by the combination of the base furnished by animal putrefaction with the vital air afforded by vegetables; observing, at the same time, that *animal substances simply, by putrefaction, afford nitrous air also.*

One of the most important facts, relative to this acid, is *its occasional existence in the atmosphere:* in establishing which we shall comprehend the several other modifications of it. And,

1. "Margraff, in the year 1751,† collected, in the suburbs of Berlin, as much snow as, when melted, afforded him one hundred measures of water, each measure containing 36 ounces. He collected the snow in an open situation, in clear glass vessels, after the atmosphere was purified by its having snowed some time: in short, he used every possible precaution to procure the snow free from every extraneous impurity. The nitrous or other salts contained in the snow, as it fell upon the ground, were no doubt dissolved in the water after the snow was melted; and in order to ascertain their quantity and quality, nothing more was requisite than to dissipate, by evaporation or distillation, the water in which they were dissolved. From these hundred measures of snow-water, he obtained, by distilling it with every suitable attention, 60 grains, not of nitre, but of *calcarious earth*, together with some grains (he does not mention the exact number) of the acid of sea-salt, impregnated with *nitrous vapour*."

2. "The same quantity of rain-water, collected in the winter months with equal precaution, and distilled with equal attention, yielded

* Townsend's Journey in Spain, vol. iii. p. 91.

† See Watson's Chemical Essays, vol. ii. p. 79, &c.

yielded 100 grains, not of nitre, but of *calcarious earth*, with some grains of the *acids of nitre* and sea-salt."

3. The same chemist found some of this acid in the common well-water of the city of Berlin; and Dr. Black, of Edinburgh, is of the opinion, that it will be found to exist more or less in the wells of all other great cities.*

4. It is satisfactory to the establishing the existence of this acid in an aeriform state, to inquire whence the plastered *lime*-walls of old buildings should acquire so much of this septous (nitrous) acid for œconomical uses: and from considering the height of such walls, and knowing the elective attraction which exists between this acid and lime, we can scarcely divert the conclusion that it is derived from the atmosphere of such habitations, and entering into combination with the lime, forms the septate (nitrate) of lime.†

5. But were all these facts deemed inconclusive, we might adduce the direct experiment of Mr. Thouvenel, who, a few years since, obtained the prize, from the Parisian academy, for *procuring nitrous acid from atmospheric air* and putrid vapour.‡

6. It was related by the professor of chemistry, from a respectable authority,§ that *water* which had been accidentally set aside in an open vessel in the Island of St. Croix, after some time became tinctured with nitrous acid, which the observer supposed to be derived from the air. Here then should seem to be a spontaneous process for its formation. And in the experiment for making water by synthesis, Mr. Cavendish constantly found that the presence of azotic (septous) gas, when the oxygene and hydrogene airs were burnt together, by the intervention of the electric spark, formed nitrous (septous) acid.|| Now, if the electric spark can affect this artificially, may it not fairly be presumed that the same operation goes on upon a larger scale naturally in our atmosphere, by the agency of lightning. Priestley also obtained an acid in the same way, as appears from his experiment relative to the decomposition of dephlogisticated and inflammable airs.¶

The further combinations of this acid, with different bases, are expressed in the Nomenclature of the French academicians, to which a reference may be had; for, considering the preceding modifications as those only which will be of immediate import to our future considerations, we shall avoid a further enumeration of them;

* Mitchill's Manuscript Notes from Dr. Black's Lectures.

† See Appendix, note A.

‡ See Bergman's Dissertation on Elective Attraction, note to page 96.

§ Mr. Van Rohr, a celebrated botanist.

|| Chaptal, vol. i. p. 229.

¶ Page 67.

them; and after observing that the septic (nitric) acid differs from the septous alone, in the superior degree of oxygenation; the septous, in its common form, being found by Lavoisier * to contain nearly two parts of septon, less than two of oxygene, and about thirteen of water;† and the septic differing from it chiefly in being more highly charged with oxygene; and that it has its respective combinations in common with it; we pass onward to state, in a more clear and satisfactory manner, the discrimination between the septous oxyd and atmospheric air, to which a separate section is reserved.

SECT. V.—*The difference between the gaseous Oxyd of Septon, and atmospheric Air.*

THEY both consist of like ingredients, but vary in their proportions and combinations. The atmosphere is composed of 73 parts of septous gas, and 27 of oxygenous; but these are mixed together *mechanically*, just as wheat and rye are distributed in a heap of grain, or as sand and shells lie dispersed along the sea-shore. The particles of each are fully mingled with those of the other; but during all this intercourse, both septon and oxygene retain their attraction for caloric in full force; and while this continues to be the case, no union takes place between the *putrifying* and *acidifying* principle, as, in common circumstances, both possess a stronger attraction for the matter of heat than for each other. The proportions of these ingredients, in our atmosphere, are not, however, invariable; for sometimes the one, and sometimes the other predominates, according as the one is absorbed or fixed, and the other effused, or set loose in greater quantity in the operations of nature. The proportions stated seem to be about the mean ratio; and the reason of their not more frequently combining and spoiling the respirability of the atmosphere, is, that when once they assume the form of gas, they have less affinity to each other than to the fire, which gives them their permanent elasticity:—The requisite to their chemical union then is, *the abstraction of their caloric.*

If there is any process in which the principle of acidity and of putrefaction lose their quota of heat, and enter into combinations with

* Chaptal, vol. i. p. 227.

† It must be remarked, that in all these compounds, in each of the forms and varieties, a certain portion of water is in ordinary circumstances always present, not as a necessary or constituent ingredient; but from the very strong attraction between it and aeriform and liquid substances, it is almost impossible to deprive them of it completely. Hence, in estimating the properties and effects of each of them, it must be ever remembered, that water enters into the composition.

with other elements, to form organized or other bodies of complex structure; in that case, the main impediment to their union, caloric in excess, will be removed, and there will be a possibility of their junction. It is admitted, that septon and oxygene enter into the composition of certain animal and vegetable substances: but, in the living state of these, they are connected with other elements, in triple and quadruple alliances, and continue thus united until their connection is severed by the disorganization of their fabric after death.

There appears thus a situation, during the decay of organized bodies, in which naked septon and oxygene may come within the sphere of each other's attraction, without entering into an intermediate state of gas, by union with caloric: and in this way, it is probable, the nitrous acid is formed during putrefaction. The process is as intelligible and easy to be conceived, as any instance of chemical connection whatever. The case, at the same time, is not by any means peculiar or unparalleled; for the union of carbone with oxygene, to form fixed air (carbonic acid gas) in the lungs, during respiration, and of hydrogene with oxygene, to constitute water in the same function, is brought about in an analogous way: and Professor Mitchill supposes it is in this way that *animal and vegetable secretions are effected*, as there appears to be some machinery in their œconomy whereby the ordinary impediments to chemical union are removed, and compounds afforded thus that are producable in no other way. This he suspects to be chiefly effected *by depriving oxygene, hydrogene, and septon, of the proportion of caloric necessary to convert them to gases*; and thus bringing the uncombined elements into closer connection, with stronger appetites to adhere to other substances and to one another; and so, most probably, gum, mucus, bile, resin, pus, sugar, starch, &c. are formed.

Hence it is apparent, that both during the life and after the death of organized beings, compounds are made that are not imitable by any art, or to be produced any where else, or by any other means, owing to the singularity of the circumstances in which the component elements then and there exist. Among these animal products, the septous (nitrous) acid, and the other combinations of septon, with the principle of acidity, have not generally been viewed in a proper point of light, either as respects their origin or properties. The alkaline quality was always thought to characterize putrefaction; and the exhalations from corrupting animal substances have ever been deemed to be some modification of ammoniacal gas. The extrication of this species of gas, in common cases, though it appears highly improbable, may, however, possi-

bly

bly take place; but even upon the supposition that ammoniac should be formed, yet the extrication of it, as will be shewn hereafter, would appear totally inadequate to the explanation of the phænomena ascribed to it.

It is only said, in addition to this, there is another compound, formed, at the same time, by the union of the principle of acidity with the same radical, and that, according to the proportion in which the former connects itself with the latter, will septous oxyd, septic gas, septous acid, and septic acid, be produced. The viscera and muscles of carnivorous and graminivorous animals contribute eminently to the formation of these; while, in the livers of fishes, and some other species of animal machinery, putrefaction disengages septon by itself, which, joining with caloric, constitutes *mere septous (azotic) gas*,* as another occasional result of this kind of disorganization. According to the texture and composition of the animal or vegetable material, will the result of putrefaction be; the septon, in the simplest case, barely uniting with the matter of heat into septous (azotic) gas; or, in other instances, combining with hydrogene into volatile alkali, or joining, under yet other circumstances, with oxygene, in varied proportions, to form the several enumerated products.

It is stated, on the authority of the Dutch chemists,† that the septous (azotic) oxyd consists of 63 parts of (septon) united with 37 parts of oxygene. The proportion of the principle of acidity is greater *in this chemical compound*, than in the atmosphere, where it is *wholly disengaged and separate*. How then, it has been asked, can *the base of vital air, the very principle of animation*, become so very deleterious as to excite distemper and destroy life, by merely combining in a small over proportion with septon? Just in the same way, it may be replied, that, by combining in a somewhat larger proportion, it constitutes, with the same radical, the more caustic and destructive septous (nitrous) acid itself. There is no reasoning *a priori*, from the known qualities of two substances, in their distinct state, to what will be the qualities of the *tertium quid* they compose by union. It may be inquired, if these two elements are so prone to unite after death, why, as they are in constant neighbourhood in the animal solid, they do not associate during life? It is probable in certain morbid cases they do; and cancer, and some other eroding and incurable ulcers of the lungs, face, neck,

* Mitchill found by experiment, in the autumn of 1795, that the fluid distending the abdomen of a swine putrifying after being strangled, was a mixture of septous (azotic or phlogisticated) and carbonic acid (fixed) airs.

† Deiman, Trooftwick, &c.

neck, uterus, &c. have been suggested as examples of its causticity when thus produced.

Notwithstanding there are but four stages of combination discriminated betwixt septon and oxygene, yet it is not unlikely, that between each there are intermediate degrees of connection; that, for instance, a given compound, divided into 100 parts, may have 99 of septon and 1 of oxygene, or 98 of septon and 2 of oxygene, and so on, or *vice versa*, in all the possible varieties of combination. This enlarged idea of the subject will go a great way toward the explanation of the *degrees of poisonous activity* in different cases of contagion; the materials of which, though always of the *same kind*, may vary in their proportions, and impart to *that mischievous fluid* more or less virulence and activity.

According to this doctrine, septous acid ought to be capable of resolution into atmospheric air. In the explosion of gun-powder this almost happens, for during the inflammation the salt-petre is decompounded; and, while the septon of its acid is set loose, the oxygene joins the charcoal to form fixed air. It is not impossible that in some instances of imperfect explosion, the septous oxyd may be formed in this way.

When putrifying substances in cities and marshes exhale this dangerous fluid, there is commonly a considerable degree of heat prevalent. In the city of New-York, the common range has been stated to be between 75 and 85 degrees of Farenheit's scale; but these fluids may be, and often are, emitted in cooler temperatures; and where the decaying materials are moist and plentiful enough, the gaseous oxyd may continue to rise in the highest temperatures we experience in our latitude of 40 deg. 40 min. What has been thus stated, is therefore to be considered as only the ordinary range in common seasons, wherein this species of air is produced from the remains of dead bodies exposed to the vicissitudes of the weather.

As to its production within doors, an opinion is entertained that the human body, whose temperature is nearly the same in every season, may produce it during the coldest part of winter; and in ships, prisons, and in some tenements of large cities, the process seems to be constantly going on. The air so formed does not seem, however, to gain its utmost activity until the arrival of hot weather, *no degree of cold we are yet acquainted with being capable of fixing it, tho' ice very effectually stops its further formation from external sources.*

Hence it is hoped the difference between the *contagious fluid* and *atmospheric air* will be apparent; the one being a *chemical connection*, whereby both the ingredients lose their separate qualities; while the other is a *mere distribution, or dispersion of the particles of each among those of the other*, possessing still their discriminating qualities,

lities, and totally free from any combination with each other: the former brought about by reason of the near approximation or stronger attraction of the ingredients in the decaying body, before they combine with coloric enough to turn them to gases; while in the latter, such is the cohesive power by which septon and oxygene, when once, as gases, united to caloric, stick to it, that the weaker attraction they have for each other cannot in ordinary circumstances overcome it; and thus, in *the same temperature*, the elements may form septous oxyd or not, according to their respective relations to other elements at the time, and the strength of alliance by which they are bound to them. If the quantity of oxygene is greater in the oxyd than in the atmosphere, it is not more surprising than in the example of arsenic, quicksilver, antimony, sulphur, phosphorus, and charcoal, acquiring great change of qualities, and becoming vastly more active by being charged with a sufficient dose of oxygene; for, like several of the enumerated substances, septon is both an *oxydable* and *acidifyable base*. That such a precise proportion of the base of vital air should be necessary to convert septon into an active poison, is not hard to comprehend, since, without it, septon is little better than a poison; it possesses no salubrious properties, and, at best, deserves but the negative character of not being mischievous.

What has been now offered is sufficient to shew the nature and theory of these septous and septic compounds, and to make it appear that the salubrity of the atmosphere, whose component parts are commonly disjoined or distinct, is spoiled immediately on their forming a junction, or becoming *chemically* combined.

This conclusion is confirmed by the following observations of Dr. Beddoes:* " The nice balance of attraction between the two constituent parts of the atmosphere deserves notice. These two substances, when closely united, form nitrous acid. If, therefore, they were not by some circumstances, prevented from uniting closely, all the oxygene, with part of the azote, would be changed into a highly concentrated acid, and the waters of our globe would be converted into *aqua fortis*," (septous acid).

And this is a perfect fulfilment of what Fontana† conjectured long ago· " Thus then the respirable and wholesome air of the atmosphere is composed of nitrous acid alone: but it is united with more or less phlogiston, which varies the quality of it, not only in proportion to its quantity, but also in proportion to its combination; because, after all, phlogiston itself is a compound of several principles,

* Considerations on the medicinal use and on the production of Factitious Airs, p. 18.

† Recherches Physiques, p. 165.

principles, presenting itself to us under a thousand different aspects, and which, combined variously with other substances, forms all the factitious airs with which we are acquainted.

"If the nitrous acid is convertable into air, and if the atmosphere is composed of this acid, decomposed, and deprived of its natural phlogiston, it might happen, in this view of the subject, that common air, in undergoing decomposition, should change to the nitrous acid, and might even form salt-petre itself: it is certain that nitre cannot be produced but in open air, that it is not formed but in places where phlogiston abounds, and that it is a salt highly charged with phlogiston."

SECT. VI.—*The Identity of what has been termed Contagion, and Marsh Miasmata.*

IT is known that onions, plants of the tetradynamous class, the paste or glutinous part of bread, &c. afford some results upon analysis greatly resembling those which animal substances yield. When natural productions of these sorts go into spontaneous decay, it may be expected they will yield aeriform products, nearly allied to the animal class, if not quite the like. Maize and rice, consisting almost wholly of amylaceous matter, emit very little or no azote, and ought therefore to be most wholesome articles of diet. The flour afforded by wheat, rye, and buck-wheat, contains a pretty large proportion of glutinous matter, abounding with azote, and should, on that account, be more prone to excite uneasiness in the stomach and bowels, by running, in certain cases, into a putrifactive state. Cabbage, onions, mushrooms, &c. should also, when taken as food, be followed by much the same consequences as meats. It is very difficult to draw the discriminating line between animal and vegetable nature; and it is uncertain whether the animal resemblances may not extend much farther than the *Fungi*, *Gramina*, and *Tetradynamia*. So far as plants or their parts approach in their nature and composition to animals, will they be capable of resolution into similar products. The difficulty started concerning vegetable putrefaction, as affording results essentially varying from animal, is so far explained away, and reconciled to fact.

The favourers of this distinction, it should seem, then, have adopted their opinion too hastily; having never ascertained that what they call *marsh miasmata* were the result of pure and unmixed vegetable putrefaction. In declaring that it was so, they have set up an hypothesis, and assumed a principle without proof, or even probability to support it. I shall take, as instances of countries

tries exposed to these vegetable miasmata, Bengal, on both sides of the Ganges, Louisiana, overflowed by the Missisippi, and Egypt, covered annually by the Nile. These tracts of country, after the evaporation, or withdrawing of a considerable part of the waters, are found to have an unhealthy air. This unwholesomeness is not imagined to proceed from any alteration in the existing atmospheric ingredients: nor in that natural distillation called evaporation, is it imagined that the atmosphere elevates, by chemical attraction, any thing but pure water. As long as this continues to be the case, there is no uncommon complaint of sickness. But by and by the mud begins to be bare, and the air grows pestilential. Hence it is said, proceeds the *marsh miasma*, from the mud and slime. Let us then inquire what this mud or deposition from the water is. Is it clear vegetable matter?—It is known to naturalists, that the *species* of large animals are comparatively *few*. It is likewise known, that their *numbers* are comparatively *small*. Nor is it less evident creation teems more particularly with animal existence in the warmer latitudes. The myriads of wild bees, locusts, ants, cockroaches, sand-flies, musquitoes, which travellers have noticed, and hundreds of other *insect species*, of which they knew neither history nor name, must, by their annual deaths, make an incalculable mass of animal putrefaction. The frogs, newts, lizards, alligators, and other amphibious creatures, which yearly expire, add greatly to the heap. The different verminous kinds, inhabitants of water and mud, contribute mightily to the sum. And to all this must be added the vast amount of fishes which die natural deaths, or are left to perish or rot upon the shores. The remains of all these animal productions are mingled with the waters of these large rivers, as they wash and cleanse the countries they pass through, from their sources in the mountains to their disemboguement in the ocean; and yet, while such is the history of the mud left by the receding rivers, and such the reason of its extraordinary fertility, the effluvia arising from it has been considered as something different from animal exhalation.

The case of stagnating water in ponds and swamps is precisely similar, abounding in animal productions, which undergo putrefaction during the evaporation of the water; and the greater heat the surface of the mud acquires, when that process begins to abate.

The same thing occurs sometimes during the long calms ships endure off the western coast of Africa, where the relicts of animals, which float within a few fathoms of the surface, now rise to the top, and form a putrid scum on the ocean water.

It is, indeed, questionable, whether there exists in nature an instance of any thing like unmixed vegetable putrefaction, upon a very

very extensive scale. Bogs of turf and of peat-moss come nearest to it; and the water issuing from thence is very palatable and wholesome. The formation of soil or *mould* in forests, is an analagous process, but not of itself particularly noxious; for wherever much morbid exhalation arises from it, there is ever an admixture of animal matter; and to this quickening of the putrefaction of manures does newly cleared land owe its abundant fruitfulness.

Sect. VII.—*Observations and Facts, proving the Identity of Cause, in the production of Fever and certain other Diseases.*

CULLEN differed in opinion from Sydenham, and thought it probable that in each of the species of disease enumerated in his Nosology, the contagion was of one specific nature.* He thought it probable that contagions were not of great variety, since they seemed to proceed much from one common source. As to miasmata, we know, says he, "only one species of it, which can be considered as the cause of fever, and from the universality of this it may be doubted if there be any other." He conceived it did not differ in kind, but varied only in the degree of its power, or perhaps as to its quantity in a given space.† Neither contagion nor miasmata were, as he imagined, of great variety.‡ And both arise from putrescent matter.§ The changing of the type of fevers, by tertians and quartans becoming quotidians; quotidians turning to remittents, and remittents altering to those of the most continued kind, appeared to Cullen to evince, not only that every fever consisted of repeated paroxysms, differing from others chiefly in the circumstances and repetition of the paroxysms; but that it was fair and proper to take a fit of pure intermittent as an example and model of the whole.‖

Balfour, in a treatise on putrid intestinal remitting fevers, comprehends the fevers called nervous, contagious, low, putrid, and malignant, together with many febrile complaints in the East-Indies, which appear with peculiar local affections. He ascribes this class of ailments to a putrified state of the *mucus lining the intestines*, which being absorbed into the blood, causes the febrile state. *This mucus* (he thinks) *receives the infection first by means of contagious matter taken into the stomach with the saliva.*¶

Wade** considers fevers and dysenteries to be ailments of a kindred nature, and prevented and cured in the same way. He considers fevers universally to arise in those latitudes, from the bowels

* Practice of Physic, Sect. 79. † Sect. 84. ‡ Sect. 85. § Sect. 86.
‖ Sect. 32. ¶ Monthly Review for July, 1794, p. 336.
** Diseases of the East-Indies, page 130.

bowels and the matters contained in them; and are to be cured of course by evacuations. He also thinks puerperal fever may always be prevented by effectual evacuations, from the bowels after delivery.

Chisholm, in his observations on the malignant pestilential fever which raged at Grenada, in the West-Indies, remarks: "Although the contagion seemed to vary much in different descriptions of people, it is highly probable that the virus of the contagion itself was uniformly the same, only variously modified by particular constitutions, habits, or mode of living, &c."

Writers have made a distinction between intermitting and remitting fevers. This may be of use in medical description and conversation; but from intermittents varying their types, and after a while becoming remittents, and then assuming the continued form, and *vice versa*, it is argued fairly that they all spring from one common cause.

Fordyce* declares, that he knows from his own observation, which is corroborated by the authority of others, that *intermitting fevers are infectious*; and this is another trait of resemblance between the two forms of fever. And Jackson † observed the changes from intermitting fever to dysentery, and from dysentery to intermitting fever, on this continent, so frequent in August and September, that he had no doubt of their dependence on the same general cause. The disease described by Zimmerman,‡ as occurring in Switzerland in 1765, shews the connection between *putrid fever* and *dysentery*. The analogy between the two diseases is very striking, and appears to indicate respiration, injured by a bad state of the atmosphere, combined with local affection of the alimentary canal. When the bowels were easy, the pulmonic organs alone were affected; the disease seems to have been what he calls the putrid pleurisy of Lousanne.

And if it has been explained, how the difficulty of animal contagion and vegetable miasmata can be got over and entirely reconciled, then *one general cause* will be acknowledged to prevail; and according to the proportion of the ingredients entering into the contagious combination; according to the sparse or concentrated state in which the product is applied; according to the part or parts of the human frame to which the application is made; or, according to the readiness or slowness with which they take upon themselves morbid action, and according to the co-operation and concurrence of other circumstances, will the effect brought on by the

* Dissertation on simple fever, p. 111.

† Fevers of Jamaica, p. 331.

‡ Treatise on the Dysentery, chap. 1—2.

the agency of this cause be. So that the recurring to a multiplicity of contagions appears both unnecessary and unphilosophical.

Thus says Merli, in his description of the contagious epidemic fever at Naples, in 1764: "This mischief, contagion, or poison, or by whatever name you are pleased to call it, produces in some a *malignant continued fever;* in others a *malignant double tertian*, or a *malignant bloody flux*. It sometimes attacks the head, at others seats itself in the breast, in the kidneys, or in some of the bowels, and wherever seated, produces the most violent and malignant symptoms."

"The disorders," says an intelligent writer,* "that prove fatal to soldiers and Europeans in general in the West-Indies, are of two kinds, namely, *fevers* and *fluxes*. They are the concomitants of armies in all parts of the world; but in tropical climates they rage with peculiar violence. There appears to be an intimate connection between them, for *they are frequently combined together, often interchange with each other, and it rarely happens that one is epidemic without the other.*" He also affirms,† that "there appears to be an intimate connection between the *intermittent* and *remitting fevers* of Jamaica: the intermittent often running into the remittent, and the remittent sometimes terminating in an intermittent." He declares further,‡ that "there subsists an intimate connection between *remitting fever* and *dysentery*; the one frequently changing into the other, and the two often complicated with various degrees of violence. In some cases the dysentery ends in a fever, though much oftener the fever terminates in a dysentery, especially among the soldiers."

"From one cause, from the same infection," says Lind,§ "I have frequently known to proceed what may be termed, from outward appearances, the *yellow petechial* and *miliary fevers*; and while, in a few, the contagion assumed an *intermitting* form, and was mild, in others it raged with a *constant* and violent fever."

"The influence of jail-infection‖ is much more extensive than is generally supposed. Of a similar nature is the disease we frequently read of in the public news-papers, under the names of the *scarlet*, *purple*, and *spotted fevers*, which often depopulates cities, and sometimes even whole countries."

Cleghorn¶ informs us that there seems to be a near alliance amongst all the succeeding diseases prevailing at Minorca. "Those who have the *rash* or *Essere* to a great degree, are very liable to *tertian fevers:* on the other hand, in the paroxysms of tertians, these cutaneous eruptions are apt to break out. The *cholera morbus*

* Hunter's Observations on the Diseases of the Army in Jamaica.

† P. 14. ‡ P. 218. § Essay on Seamen, p. 265. ‖ Ibid, p. 323. ¶ Diseases of Minorca, ch. ii.

bus sometimes hath its regular periods like a *tertian*, as the paroxysms of tertians are frequently attended with a cholera. Sometimes a *tertian* turns into a *dysentery*, or a dysentery becomes a tertian; and when one of these diseases is suppressed, the other often ensues: nor is it uncommon for *dysenteric fevers* to put on the form of *tertians*, and for the fits of tertians to be regularly accompanied by gripes and stools."

That the diseases produced by contagion and miasmata originate from the same cause, is countenanced also by Gardiner,* who coincides with the physicians of observation and experience, that marsh miasmata can acquire a power as noxious as human contagion; and when it does so, the distempers caused by it are nearly the same. Sir John Pringle was of this opinion, and so was Count Carburi. And what is more to the point, Dr. Mackenzie, who had resided along time at Smyrna, and a longer at Constantinople, declares the common epidemic pestilential fever there, is the same with the goal or hospital fever of England; and when this same distemper grows more virulent, with buboes and carbuncles, they call it plague.

Cleghorn† further asserts: "These *tertian fevers* have as good a right to be called *contagious*, as the measles, small-pox, or any other disease; for although in that season there certainly is a peculiar disposition in the air to affect numbers in the same way, yet those who are much conversant among the sick, are most liable to catch the distemper."

SECT. VIII.—*That the Cause of Contagion, and of many endemic and epidemic Diseases is some Chemical Combination of Septon with Oxygene.*

THE effluvia from putrifying substances, which constitutes contagion, is neither hydrogene gas, nor any combination of it with sulphur, charcoal, or phosphorus. These compounds are very volatile and diffusable, and form a large part of the disagreeable odour, or abominable stench of decaying bodies. The stinking smell of substances is quite a different thing from infection. Nor can carbonic acid air be the contagious material, though that exhales abundantly from some sources of corruption. It has been imagined, that ammoniacal gas was the injurious production, either by itself, or in combination with something else; but the sensible qualities of this, although it supports flame, and is miscible with water, serve sufficiently to characterize it, and shew it is not the deleterious cause in question. Besides its miscibility with water, and

* See Animal Economy, p. 187. † Diseases of Minorca, note to p. 132.

and capability to maintain flame, though very faintly, alkaline gas possesses enough of peculiar qualities to distinguish it from every other animal production. We shall mention two of them: 1. When ammoniacal gas is mixed with water, it imparts to it a strong alkaline tincture; insomuch that a water may be prepared in this way, having a stronger alkaline smell than any spirit of sal ammoniac at all. 2. Whenever alkaline air meets with carbonic acid gas, a combination of the two fluids takes place immediately, even in their aërial form, and concrete into oblong and slender chrystals, which cross each other, and cover the sides of the vessel in which the experiment is made, in the form of a net-work; the chrystals being of the same kind of volatile salt, obtained in a solid form, by the distillation of sal ammoniac with fixed alkaline salts. Hence, if ammoniacal gas is ever extricated during putrefaction, it would instantly discover itself by imparting an alkaline flavour to water; or, by combining with the fixed air, evolved at the same time, will combine into firm chrystals of volatile alkali. This, therefore, cannot constitute the matter of contagion. *This contagious cause we suppose to be sought for in the combinations of septon with the acidifying principle, and to manifest itself in the septous oxyd, and the vapours of the nitric acid itself:* and in this view of the matter can we account for the production of contagious diseases in different parts of the world, wherever the causes favouring the production of these compounds exist.

Upon this idea, the occurrence of the epidemic in this city last summer, may admit of a satisfactory solution; from the existence of the collections of vegetable and animal substances in the different divisions of the city, and in a particular manner, in that part of it where the malady raged with peculiar violence; since, on examination, it has been found, that there existed in that neighbourhood large heaps of manure, collected by the scavengers from the streets and avenues of the city, and which were in that situation, undergoing the necessary disorganization, for agricultural purposes. Now it has been pretty fully stated, that under such circumstances, those septous compounds, which are the immediate causes of contagious and infectious ailments, are uniformly extricated by the application of heat and moisture; and when once so formed, their influence on the neighbouring inhabitants is easy of conception.

An analogous influence from the marsh effluvia arising from the borders of the Onondago lake, is related by Vandervoort.*

"The marsh effluvia in this western territory, in many places, and particularly in this place, operates so powerfully on the human body,

* Analysis of Ballston mineral spring water, p. 17.

body, as to induce a paroxysm of an intermittent, in the course of four or five hours, and frequently death the seventh day.

"From ocular observations in these marshes, it appears, that the poisonous effluvia is generated from the putrefaction of vegetable matter, which, in its resolution, undergoes certain changes, which produce this noxious air. It is also evident that this air does not operate while the marshes are inundated."

Moores* informs us, that in the state of Maryland, where the marshy grounds are more extensively prevalent, remitting and intermitting fevers prevail; and both diseases he ascribes to the same cause, differing in degree. "Hoc vero ex observatione propria edidici nempe cum in Marilandia, aestivus et autumnalis calor minor quam solitus sit, tum populi in locis paludibus vicinis degentes febribus intermittentibus tantum obnoxii sunt quando autem aestatis et autumni calor intensior aestuet tunc febres remittentes his ipsis locis epidemicæ grassantur.

"In regionibus Marilandiæ calidioribus qualis Carolina est, paludes solis radiis ferme excoquuntur, ibique febres remittentes quam in Marilandia multo gravioribus symptomatibus stipantur et proprius ad typhum icterodem accedunt. Monstrant hæc exempla quantum febres cum ejusdem tum diversarum regionum secundum tempestatis calorem varient. Hæ autem varietates modo indirecte a calore pendent quippe paludum miasma pro causa omnes agnoscunt et secundum vim ejus febres remittentes vel intermittentes sæviant; vis vero miasmatis ad tempestatis calorem semper quadrat."

Dr. Valentin, who formerly resided in Cape-François, in the capacity of physician to the camps and armies of St. Domingo, and who was in Norfolk, in Virginia, during the sickness of 1795, in a letter to Professor Mitchill, has the following comparative remark on the diseases of the two places: "They offer the like train and concomitancy of symptoms; I have here followed the same method as there, with an equal success, when I was called in season. I do not contest about the word *yellow-fever*; that I consider but as an effect, or a symptom, for it is not a new malady." He adds also, his entire conviction of its local origin in Norfolk, and other sea-ports of the United States.

The contagious fluid, emitted from living bodies, is most plentifully conveyed in the breath, perspiration, and stools. It has been said to have a peculiar smell, and capable of being distinguished from all other known odours. They who have had infectious air fresh in their nostrils, have called it an earthy, disagreeable smell, affecting, in some degree, the organ of taste, and extending

* Tract. Inaugural. de Febre Remit. Marilandica.

extending down into the stomach: some have compared it to the vapours issuing from a newly opened grave, but without the cadaverous stench; others think it resembles the effluvia, of rotten straw, and others again are of opinion it is like the exhalations from confluent small-pox, at the turn of the pustules.

From the circumstances in which it is emitted, it is presumable it is seldom admitted to the organ of smell in its *pure form;* but is generally accompanied with some other gaseous emanation floating about with it. Perhaps it is impossible to obtain it in a pure form, but by an artificial process, and this may be the reason of the diversity of opinions concerning the odour ascribed to it, which is probably not so much occasioned by the contagious fluid itself, as by the other matters that are frequently extricated at the same time with it. After diffusion through the air to some distance, it seems incapable of exciting any sensation at all in the organ of smell. From this inodorous quality of it, added to its capacity to support flame, may some idea be formed why it has hitherto eluded the search of inquirers.

The facts related by Mr. Martin,* and by Mr. Townsend, concerning the vapours rising from the salt-petre soils of Bengal and Spain, and their power of producing fevers, apply with great force here. They are nitrous earths, naturally formed. The filth collected in the streets of large cities, is a nitrous soil also. The effluvia from the salt-petre soils of cities produces effects very similar to those observed in the neighbourhood of natural nitre-beds. This is verified most strikingly, as before remarked, in the disease endemic in New-York in 1795. The inference is, that the septic (nitric) vapours, according to Thouvenel's conclusion, are, in both cases, the cause of the consequent diseases.

SECT. IX.—*The Operation of the Causes of these endemic and epidemic Diseases on Vegetation.*

THIS is a subject of the greatest importance, as it involves a material article of the police of crowded cities: for should it appear that the vegetable economy was capacitated to disarm these compounds of their baneful properties, the joint co-operation of convenience and usefulness would stimulate the officers of government of these places, to disseminate vegetable life with as much zeal, as the prejudice of ignorance has supported the measure of exterminating it from the streets and public walks of the citizens.

No direct experiments, however, have yet been made on this subject; but from the phænomena of the mitigation of the Samiel and

* See Appendix, note B.

and Harmattan blasts in Africa, on their passing over tracts of country cloathed with vegetable verdure, it is highly probable they have an essential influence in altering or decomposing the elements of such pestilential fluids.*

CHAPTER II.

Its Physiological and Medical Operation.

HAVING, as we hope, ascertained the cause of most endemic and epidemic diseases, and the sources of their formation in the various ways we have considered them; we come now more immediately to the application of these causes on our bodies; and in this view of the subject we shall consider the operation of them, on different parts of the body, in separate sections.

SECT. I.—*The Operation of the Gaseous Oxyd, and Septous Acid Vapours on the Lungs, including Brute Animals as well as Man.*

THE operation of these causes on the lungs is accompanied, according to their degree of continuance, with the following symptoms:—Hoarseness, cough, catarrh, excretion of mucus from the larynx or bronchia, lassitude, languor, shivering, impeded oxygenation of the blood, the contractions of the heart diminished, intermission or slowness of the pulse, the colour of the hands and arms brown, livid or black, laxity of the muscles, hæmorrhagies, anxiety, coldness of the extremities or of the whole body, stupor, coma, delirium, suffocation, and death frequently direct on the first attack; the blood in a like condition as in submersion or suspension.—Though the effects now enumerated seldom all occur in any one case, some or other of these symptoms will occur, more or less, in different persons, dependent on the concentrated form in which the cause is applied, the duration of its continuance, and the facility with which the patient's constitution accommodates itself to the operation of such a new stimulus. And in this way it is, that of a number of individuals, labouring under the influence of causes of this sort, in like situations and circumstances, one shall have an aggravated, another a mild, and a third no disease whatever: and so on in all possible degrees of violence or mildness.

For a more clear and satisfactory illustration of these remarks, I shall select, from medical history, a few well marked and decisive instances of diseases induced by breathing an air thus vitiated.

Chisholm's

* See Appendix, Note D.

Chisholm's* history of what he calls an uncommon epidemic fever observed in the island of Grenada, may be considered as a fair exemplification of the effects produced by this modification of atmospheric air, mingled perhaps with other non-respirable airs, and acting on the lungs of the soldiers, sickening in the barracks of St. George's, a remarkable unhealthy spot, *surrounded by marshes*. " The general type of the fever was that of a quotidian intermittent, but so extremely irregular as not to admit of a reference to any of the common species. It was truly anomalous, and so insidious as to endanger the life of the patient before any apprehension could be entertained of its fatal tendency. In almost every case, the patient seemed in a state of very great anxiety at all times, with eyes inflamed, and a little protruded; a strong expression of depression of spirits in his countenance; a very great degree of debility, and a sense of weariness, as if he had undergone excessive fatigue. But the symptoms which most troubled the sick, during both the paroxysm and the intermissions, if they can be properly called such, were an intolerable headach, with a throbbing of the temples, and a lethargic heaviness. During the intermission, whilst the patient was labouring under all, or most of these symptoms, his skin was preternaturally cool; his pulse small, quick, and hard; and his whole body covered with a clammy moisture. The paroxysm generally came on sometime between eight and twelve at night; and its approach was indicated by a very great increase of the coldness, with shivering. These were soon succeeded by violent heat, increased anxiety, and headach, and very frequently by delirium. It continued two, three, or four hours, and terminated in profuse diaphoresis; but contrary to the usual form of intermittents, on the ceasing of the diaphoresis, the patient continued afflicted with anxiety, headach, &c. to the degree I have mentioned. The paroxysm, in some, was marked with infinitely greater violence than in others. In the case of one soldier, the paroxysm began about twelve o'clock at night, with all the most violent symptoms, at once a deadly coldness of the body, and excessive delirium. These, constantly increasing, terminated, in about two hours, in total insensibility, coma, and death. In a few cases, very little intermission could be perceived throughout the whole course of the disease: in these there was a continual alternate succession of shiverings and flushings, with a disagreeable clamminess on the surface of the body, which never afforded relief. When, in such cases, any thing like distinct intermissions could be observed, they occurred at or a little before noon, and continued one or two hours; but in the others, the periods of shivering and diffusion

* See Duncan's Medical Commentaries for 1792, p. 267, et seq.

diffusion of heat returned so rapidly, that scarce any interval could be perceived. The prostration of strength, brought on by them, was astonishing; and it was observed too, that the patients who laboured under the disease in this form, had a much more ghastly countenance; sighed and moaned more; were infinitely more restless; were more subject to raving, and had more of a dirty, yellow suffusion over the skin, than the rest.

"In most of the cases respiration was extremely difficult; and on ordering the patients to make a deep inspiration, they were suddenly checked by pain more or less acute, which, however, they could not refer to any particular place; except in a few instances, when it was found most troublesome at the pit of the stomach, stretching to the hypochondria and spine. Some also complained of a *rawness*, as it were, from the throat to the stomach; or, as they expressed it, "a rawness and burning of their inwards."

"A yellowness on the skin was by no means a constant symptom. The suffusion was general over the body; sometimes of a deeper hue, and sometimes particularly about the eyes, and on the neck, approaching to a livid colour.

"Dissections, in these cases, proved the whole intestinal canal to be sound."

Lind* gives an account of a disease, operating pretty much in the same way, though with less violence, on board several ships in the British navy, in 1759. "The fever which raged in all these ships, greatly affected the breast. Some who were seized with it, as if they had been under a salivation, spit up six or eight pints of a thin phlegm, in forty-eight hours; and, to prevent suffocation, were obliged to have their heads supported by pillows. Their blood was extremely viscid and glutinous. This I observed even during the last stage, in a person from whom it was then necessary to take blood for a pain in the breast, greatly impeding respiration. The head was affected often with a heaviness, and dull sense of pain, seldom with a delirium. Cough, spitting, and pricking pains of the breast, were the most universal complaints, &c. The attack of this infection begun with shivering, succeeded either by pain of the head or breast; seldom by an universal pain in the limbs, but most frequently by a tightness of the breast and cough, which last raised the acute pains in the chest. Several who recovered were afterwards distressed with a dulness of hearing; many relapsed. A midshipman, after being able to sit up for several days, fell again into the fever, which was accompanied with convulsions, and expired in thirty hours after the last attack, when his body was found covered with patechiæ: a few died

* Dissertation on Fevers and Infections, Ch. 1.

died consumptive, being exhausted by the vast discharge in spitting. In four or five persons, there were symptoms of malignity; and out of above an hundred patients received into the hospital, from those ships, eight died of the fever. The distemper, if it had occured elsewhere than in the ships, might perhaps have been judged solely inflammatory, and to have proceeded from causes very different from the real ones."

Fordyce, in his dissertation on simple fever,* has the following observations, which tend to the corroboration of this argument. "When the attack is fatal, it sometimes kills in five minutes: sometimes it requires half an hour; sometimes longer than that time. While the patient is yet sensible, violent headach, with great sense of chilliness, takes place; the extremities become cold and perfectly insensible; there is great prostration of strength, so that the patient is incapable of supporting himself in an erect posture. He becomes pale; his skin of a dirty brown, and he is soon insensible to external objects: the eyes are half open; the cornea somewhat contracted, and the patient goes off very soon: the pulse is diminished, and at last lost, without any frequency taking place; but if it be long before he dies, the pulse becomes excessively small and frequent," &c.

The effects induced by exposure to an atmosphere, charged with the causes of these diseases, as *pointedly* elucidating their operation on respiration, may be seen in Mr. Martin's account of the air of some parts of Bengal.†

It may not be amiss, in this section, to investigate *the operation and influence of these causes on brute animals.*

From the production of diseases in the human constitution, by exposure to marsh miasmata, and other like decompositions of animal and vegetable snbstances, it may appear highly probable that the *rot in sheep*, kept in low grounds, is analogous to the intermitting, and like diseases of the human species, and excited by the same causes. And a like investigation might prove that *the bloody murrain*, which prevails occasionally among *cattle*, is occasioned by a similar contagious cause, operating and producing an affection similar to our *dysenteric* fever, &c.

But that, while our habits suffer the operation of these causes in their superior degree of operation, brute animals have analogous experience, may be inferred from the following facts:—

Sorbait, of Vienna, relates, that during the plague, "*larks*, so numerous in Austria during the autumn, were wholly wanting, so that not a single one could be met with; and *tame birds*, kept in

E cages,

* Page 181.

† See Appendix, note B.

cages, *all died.* Homer* mentions the death of *dogs* and *mules*, as the forerunners of the pestilence in the Grecian camp before Troy. Thucidides† observed the pernicious and deadly effects of the atmosphere, during the plague at Athens, in the second year of the Peloponesian war, upon *birds* and *beasts*, and particularly on *dogs*.

Facts of this kind were not wanting during the prevalence of our late epidemic. It is stated to me, from undoubted authority, that *at the Belvieu hospital* the *fowls* and *chickens*, which fed about it, and like other poultry, came occasionally into the hospital, (perhaps drank, and picked up crumbs about the house,) all suddenly died without any ostensible cause; on dissection, the appearances were similar to those of the human species who have died of putrid diseases. And in a part of the city, where there existed comparatively, a small proportion of the epidemic influence, the *ducks* kept in yards, and which received the usual attendance, as to food, all died off in such a manner, as to excite the remarks and surprize of the inhabitants.

Fowls, in some of the yards, sickened, and some died; but after a while, the survivors experienced no inconvenience. Bruce mentions the deadly effects of the Simoom blasts, in the African deserts, upon all breathing creatures. Cleghorn observes, that in some of the valleys of Minorca, beasts, as well as men, suffer; and it is related by respectable authority, that *horses* have been incommoded by the autumnal air of Onondaga, for some time after being carried on to the low lands.

SECT. II.—*On the Alimentary Canal, or Stomach, and Intestines.*

How the aforesaid Septous Combinations may be conveyed into the Alimentary Canal of living Animals; or generated there from the Corruption of the Food; and of the Effects they produce in both Cases, by acting on the Stomach and Intestines.

1. FROM the properties of the causes of contagious and infectious diseases, then, we infer the occasional operation of these in the alimentary canal, by their introduction *by the saliva and watery parts of our food:* in corroboration of which mode of operation, we are presented with the following reflections and facts from their respectable authors:—

Gardiner observes that the manner in which the causes of certain malignant fevers, arising from marsh miasmata, human or specific contagion, get into our habit, is by the noxious effluvia taken in with the air in respiration, mixing with the *saliva*, and,

* Iliad. i, 69.
† Book ii.

and, by deglutition, conveyed into the stomach, where by certain changes wrought on the gastric fluids, and their particular stimulus on the nerves of the stomach and bowels, they prove the cause of fevers, differing from one another according to the nature of the infectious exhalation. "This I have always thought to be the most probable way that infections are received by us, and of their acting on our system, in the production of fevers."*

This opinion of the miscibility of the infectious effluvia of malignant fevers, with the *saliva*, and of its afterwards passing into the stomach, is upheld by Turner, physician to the military hospitals in the West-Indies, in a letter to Sir John Pringle, who says that he escaped the infection of the hospital fevers by chewing tobacco during the time he was on his visits to the men in the hospital, imagining that all putrid and contagious effluvia entered and infected by the *saliva*, which he took great care not to swallow whilst he visited the sick.†

Lind believed in this mode of receiving contagion, who says, "Swallowing the spittle in infected places is justly deemed a means of sooner acquiring the taint, upon which account neither the nurses nor any one else should be suffered to eat in the hospital."‡ And he relates an occurrence which may unquestionably be instanced as operating in this manner—"A company of gentlemen belonging to his Majesty's ship the Phœnix, taking the diversion of hunting and shooting at the mouth of the river Gambia, by following their game into a large swamp, were all of them affected by its putrid effluvia. They were immediately siezed with sickness, vomiting, headach, and a constant hawking and spitting from a disagreeable smell which (as they express it) seemed to remain in their mouth and throat. Upon returning to the ship, each of them was ordered a vomit, which immediately removed all those complaints."§

And again,‖ "I am apt to think, that an infection from whatever impure fountain it is derived, does first discover itself by affecting the stomach and intestines."

Balfour's opinion corroborates this argument, as we have seen in our collection of observations and facts proving the identity of cause in the production of fever, and certain other diseases.¶

Rush

* Animal Economy, p. 196.
† Med. Annot. vol. v. p. 472.
‡ On Hot Climates, p. 111.
§ On Hot Climates, p. 138.
‖ On Fevers and Infections, part 2, p. 65.
¶ See p. 23.

Rush* relates, that "in small rooms, crowded, in some instances, with four or five sick people, there was an effluvia that produced giddiness, sickness at stomach, a weakness of the limbs, faintness, and, in some cases, a diarrhœa."

Mitchill relates an instance of a gentleman who received the fumes of a dysenteric purging: some uneasiness of the stomach, and qualmishness, came on, which, in a few hours, ended in a *looseness*. A similar indisposition was brought on in another, by receiving the fumes of a corrupting corpse. In both these instances, it appears as if the infection was swallowed with *the saliva*, and thus operated as a purge.

Mr. Van Eems† informs us, that if the abdomen of an animal that has laid some time under water after drowning, suddenly burst, a most active and penetrating fluid proceeds from it, very injurious to the eyes, lungs, and stomach; and instantly destructive to the appetite, insomuch as to provoke nausea, vomiting, and even to bring on dysentery.

2. Another mode in which the diseases dependent on those causes may be produced, is *by the disorganizing process, which the ingesta take on, in the stomach and bowels themselves.*

The human constitution derives its principal support from foreign materials, received into the digestive organs: and the due performance of this necessary function is therefore indispensably requisite to the healthy condition of our bodies: this state is preserved by the influence of the gastric liquor and saliva in the stomach, and by the pancreatic and bilious juices in the small intestines; but these healthy secretions are again dependent on the vigour of the organs of digestion, which is increased or diminished by the matters taken into them. These substances are derived from the animal and vegetable kingdoms, and in healthy digestion, will they sustain and nourish our systems.

But in cases of the prevalence of epidemic, and other like diseases, while the predisposing causes exist, it would be to be inferred, from our principle, that flesh-eaters, in an especial degree, should be liable to an attack of those diseases; for, from the analysis of the lean muscular fibre, we are enabled to understand whence may flow the septon, for the formation of the deleterious cause; and by the occasional use of vegetable substances, as well as from the animal matter itself; the other ingredient, oxygene, is afforded. A consequence is, that those substances which are deficient in either of the component ingredients, are those of the most salutary use in such cases: substances of this nature are the

* On the Remitting Bilious Fever of Philadelphia, 1793, p. 107.
† Præ ection. Boerhaav. 248.

the fat of animals, oil, butter, &c. which, are known not to contain the septous or putrifying base, and may therefore be had recourse to as food, with the most perfect safety; or, on the other hand, vegetables, as affording less septon than lean meats, and some of them scarcely any, may be used with like freedom. And it is unquestionably owing to this mode of living, that Frenchmen are comparatively in so small a degree, sufferers from these forms of pestilential ailment, in unhealthy countries and climates; whose habits of life are such, as to draw by far the greater part of their nourishment from vegetable substances. To this effect, Jackson* observes, that "the French and Spaniards eat less animal food, and drink their liquors greatly more diluted, than the natives of England: they escape better from dangerous diseases, and this has been remarked to bear some proportion to the degree of abstemiousness they observe." Mr. Verdoni declares, that "the Greek christians in Smyrna, during the season in Lent, when they eat only vegetables, are very seldom attacked by the plague, while among those who eat flesh the contagion makes great havock. Thence the best means of prevention are to eat moderately, and not at all of animal food," &c.

In infected situations, those causes which tend to the debility of the digestive organs, will contribute very much to the production of disease. Of this kind are the habitual use of too much ardent spirits, and other habits of intemperance; exposure to excessive heat, and other violent exciting powers, which, by their operation, induce an indirect debility of these organs; or the operation of fear, terror, and the like, may induce a similar operation, by an abstraction of the usual necessary stimulation. These favour the unhealthy disorganization of the ingesta, by withholding the antisceptic powers of digestion and concoction, depending upon a healthy state of the gastric secretion. When, by the operation of the above-mentioned noxious causes, this unhealthy and debilitated condition is induced, it is evidenced to us by tension, oppression, and belching; and still further, by the acid, putrid, and pestilential nature of the matters vomited from the stomach.

That an acid is vomited up in cases of this nature, is ascertained by the sour taste, and by the erosion of the calcarious enamel of the teeth, in some instances. "I have," says Hunter,† "seen an instance of fever, in which it was necessary to give from half an ounce, to six drachms of the powder of oyster-shells, to destroy the acid that was generated in the course of the day, which otherwise

* Fevers of Jamaica, p. 395.

† Observations on the Diseases of the Army in Jamaica, p. 161.

wise occasioned great pains and retchings." The green colour of the bile is known to depend upon an acid in the stomach; for we know by experiment, that the most healthy bile, whose colour is yellowish, will be changed to green by mixture, in certain circumstances, with an acid.

What the nature of this acid is may be collected from the account given by Wallis,* of a porraceous and black bile often seen thrown up by vomiting, which corroded metals, and boiled up on the ground like spirit of vitriol dropped thereon; and so austerely acid, that it set the teeth strongly on edge, and excoriated the throat. It is further related, that a man who had vomited up a large quantity of green, black, and acid bile, being agitated by convulsions, had a silver spoon put into his mouth, that he might not bite his tongue, and in a moment it turned black as if it had been stained with spirits of nitre; but the man had drank liquors soured with lemon juice.

The coffee coloured matter, commonly called the *black vomit*, which is ejected from the stomach in violent remitting fevers and plague, is to be considered as bile, impregnated with the more acive septous combinations, as appears from its corrosive nature, noticed by dissectors in these diseases; or yet, in other cases, there is an admixture of blood, while by its own caustic nature the septous acid thus formed in the stomach, had eroded the extreme blood-vessels of that viscus. Hillary† witnessed the existence of these appearances in the remitting fevers of Barbadoes, who observes, that "great quantities of black, half-baked, or half-mortified blood, are frequently voided, both by vomiting and stool; with great quantities of yellow and blackish putrid bile, by the same ways."

When, therefore, there exist such fluids in the stomach, the occurrence of nausea, excessive vomiting, painful burning of the stomach, and other symptoms of Gastritis, we should scarcely doubt from the nature of these septous combinations, that on dissection, the various phænomena of inflammation would manifest themselves.

Conformably we find, according to the dissections of Drs. Physick and Cathrall, that in the disease of Philadelphia, in 1793, "*the stomach and beginning of the duodenum, are the parts that appear most diseased.* In two persons who died of the disease on the fifth day, the villous membrane of the stomach, especially about its smaller end, was found highly inflamed: and this inflammation extended through the pylorus into the duodenum some way. The

* Edition of Sydenham, vol. ii. p. 191.

† Diseases of Barbadoes, p. 151.

The inflammation here was exactly similar to that induced in the stomach by acrid poisons; as by arsenic, which we have once had an opportunity of seeing in a person destroyed by it.

"The bile in the gall-bladder was quite of its natural colour, though very viscid.

"In another person who died on the eighth day of the disease, several spots of extravasation were discovered between the membranes, particularly about the smaller end of the stomach, the inflammation of which had considerably abated. Pus was seen in the beginning of the duodenum, and the villous membrane, at this part was thickened.

"In two other persons, who died at a more advanced period of the disease, the stomach appeared spotted in many places, with extravasations, and the inflammation disappeared. It contained, as did also the intestines, a black liquor, which had been vomited and purged before death. This black liquor appears clearly to be an altered secretion from the liver; for a fluid, in all respects of the same qualities, was found in the gall-bladder. This liquor was so acrid, that it induced considerable inflammation, and swelling on the operator's hands which remained some days. The villous membrane of the intestines in these last two bodies, was found inflamed in several places."

And in a letter addressed to Dr. Duffield, of Philadelphia, from Professor Smith, we extract the following appearances on dissection, of one who died of our epidemic, in 1795: "*The stomach and duodenum exhibited marks of high inflammation;* the liver little altered from its natural appearance: the gall-bladder absolutely empty; the lower intestines quite sound, the lungs also: and the brain, except in being rather of a darker shade than is usual."

From these dissections it appears, that the superior portion of the intestinal tract, is the part which more particularly suffers the operation of these causes of disease. And hence we are led to an inquiry into the causes which may account for this partial operation. From the nature of these septous combinations, which are supposed to be the offending causes in these situations, we are naturally led to inquire into the component ingredients of the bile, which is here emptied into the intestines; in order to ascertain if there does not exist in it some substance, for which the principle of putrefaction has a greater attraction than for the viscera themselves.

From the succeeding experiment of Saunders,* we analogically conclude, that such a substance does exist in the biliary secretion of the human species; as there is probably no very material difference in us, from other animals, in this secreted liquor.

"A quantity

* Treatise on the Liver, p. 103. London, 1793.

"A quantity of *(ox)* bile and diluted marine acid were put into a flask, and placed in a sand bath, until they had acquired the boiling heat. On inspection, the separation into parts was very evident; and on committing it to the filter, it separated a colourless fluid, destitute of every bilious property. The *residuum* consisted of a very dark green mass, intensely bitter, and extremely glutinous. When examined, it appeared to be composed of an animal mucilage, in combination with a resinous substance.

"But to ascertain in what way the acid had effected the decomposition, it became necessary to examine the filtrated liquor. It was therefore subjected to a cautious evaporation, and at a proper period was suffered to cool.

"Under cooling, chrystals were formed of a cubic figure, which decripitated by heat, and possessed all the characters of common salt.

"Therefore the decomposition was here occasioned by the marine acid engaging the *mineral alkali*, which it separated from the other elements of the saponaceous body, and, by uniting with that *basis*, formed common salt."

Since then there is existing in our intestines, this bilious secretion, whose alkaline nature we have thus seen experimentally ascertained, it will appear manifest from the known attraction between the acid of the stomach and the soda of the bile, in what manner there shall take place a union of the septous acid with the biliary alkali, and thus a considerable part, or in some cases even the whole of the septous acid, being thus neutralized or saturated, that part of the duodenum below the opening of the ductus communis choledocus, together with the jejunum and ilium, in most instances, be unmolested by its inflammatory and caustic nature. These facts lead to a satisfactory solution of the phænomena of the extraordinary secretion of the bilious matter in those diseases; since it appears, from the preceding considerations, to be one of the resources of nature, in counteracting this offending cause.

Or, if it should not meet with a substance to neutralize it wholly, or nearly so, it may pass on, and by the operation of the absorbent vessels, by being taken up and carried the round of the circulating fluids, be eventually manifested by imparting to the skin a yellow, or other like suffusion: this explication of the tinge of the skin, in certain instances, is rendered probable, by the relation of a case of disease last summer, from undoubted authority, in which the following striking fact presented itself: A patient labouring under the epidemic, whose skin was of a remarkably yellow colour, was ordered an epispastic; on dressing the blister

ter after its operation had ceased, the scissars which were made use of to remove the scarf skin, were accidentally left moistened with the serum for about a quarter of an hour, when, on examining them, the observer was astonished to find they were oxydated as if dipped into a strong mineral acid.

And Van Swieten, in the plague of Oczakow, relates, that "the instruments which the surgeons made use of, were turned as black and livid as if they had been dipped in aqua-fortis."

Sect. III.—*On the Cuticular Surface.*

How the Skin of Persons, living in an Atmosphere thus vitiated, is affected, and how it operates on Wounds.

ON the application of the causes of disease which we have been considering, do we presume, are to be explained various efflorescences and eruptions, occasionally besetting the external surface of our bodies. These appearances will be different, according to the concentrated or sparse form in which the causes are applied, and to the duration of their continuance; and in this way partly is to be explained the peculiar tinges of the countenance in tropical climates; even although there shall not be present actual disease.* And, on the like principle, is explicable the absence of the different suffusions in contagious diseases, till some progress has been made in the morbid action.

To this mode of operation, are probably referable the different kinds of eruptions, which have been hitherto considered as critical depositions of humours from the body. From the greater specific gravity of the contagious fluids, the lower parts of the body, which are most constantly exposed to its operation, are first affected, and worse afflicted than other parts of the constitution. Thus the legs of persons sick on ship-board, are often miserably ulcerated. Sometimes, in very unhealthy countries, an uneasy itching in the legs, has been the first symptom of disease, and upon pulling down the stockings, streams of thin dissolved blood followed; soon after which a ghastly yellowness invaded the whole body, and the patient has died in less than forty-eight hours.

In 1764, the quality of the air at Batavia, which then exhibited a mingled scene of disease and death, was so malignant, that a slight cut of the skin, a scratch of a nail, or the most trifling wound, &c. generated quickly into a putrid spreading ulcer, which, in twenty-four hours, consumed the flesh even to the bone, as was experienced by the British frigates Medway and Panther, then lying there. From its disposition to adhere to bedding and clothing, there

* See Appendix, note E.

there can be no difficulty in understanding how miliary and petechial spots are produced, in many instances beginning about the back, loins, and inside of the thighs of those who are exposed, and extending thence over the covered parts of the body, in those who are sweating in their own vapours and exhalations in bed: they have thus a contaminated atmosphere around them; and, from its operation upon the sick, do these morbid appearances of the skin proceed. The occurrence of these sorts of fevers in low, foul, and dirty dwellings, &c. may thus be referred to the noxious air the patient lives in, and to the infected beds they lie upon.

In the island of Jamaica* sores are very frequent and troublesome on the lower extremities. A trifling scratch, bruise, or hurt on the feet and legs, soon became a deep and spreading ulcer, which was always difficult, and sometimes impossible to be healed. Little injuries of these parts, are very apt to spread rapidly, and form a large ulcerated surface. Granulations are hard to form, and when formed, frequently become flaccid and mortify, the portion skinned over ulcerates afresh, and the sore becomes larger than ever. The bones are apt to become carious, the patient to grow hectic, and linger on to death. Fresh vegetable diet, full nourishing diet, calomel in smaller doses, to operate as an alterative, external application of almost all sorts of poultices, ointment, dry lint, &c. and a horizontal posture of the limbs, were so ineffectual, that Hunter relates it as the general result of all his experience, that ulcers of some standing, and of considerable size, in the lower extremities, cannot be healed in that country, by any means we are acquainted with. Yet these very patients sent away from the island to Europe, had their ulcers frequently healed on the passage; the rest, except in cases of carious bones soon recovered. Amputation in these other cases answered in England, but succeeded very badly in Jamaica.

Cleghorn† testifies to the analogy between Rome and Minorca, in respect to the troublesome nature of ulcers, on the inferior extremities. "Baglivi tells us, that at Rome ulcers of the legs are almost incurable, and wounds in them difficult to heal, while the like accidents on the head are quickly cured without any trouble. The same thing happens here, insomuch that it is a proverb among the natives, "Minorca is good for the head, but bad for the shins."

Van Helmont‡ saw a man, who, upon touching some papers infected by the plague, felt instantly a pain, like the prick of a needle;

* Hunter on the Diseases of the Army in Jamaica, p. 227.
† Diseases of Minorca, p. 78.
‡ Tum. Pest. p. 853.

dle; a pestilential carbuncle made its appearance soon after, on his fore finger, and he died in two days.

Van Swieten* relates, that a man who stirred up with his foot the straw whereon the bed of one sick of the plague had been laid, "a little while after felt an acute pain in the lower part of his leg, just above the foot, as if the part had been scalded with boiling water; the next day the epidermis, or scarf-skin, was elevated into a large blister, upon breaking which a quantity of blackish liquor run out, and underneath a latent pestilential carbuncle was discovered."

SECT. IV.—*On the Lymphatic and Glandular Systems.*

How the above-mentioned Compounds operate upon the absorbent Vessels, and the Glands of the Body.

ON these parts of our systems we are presented with the phænomena occasionally attendant on dissections of dead bodies, from the peculiar gas thence arising, and of which an account is given by Mr. St. John.† "I have known a gentleman, who, by slightly touching the intestines of the human body, beginning to liberate this corrosive gas, was affected with a violent inflammation, which in a very short space of time extended up almost the entire of his arm, producing an extensive ulcer of the most foul and frightful appearance, which continued for several months, and reduced him to a miserable state of emaciation," &c.

A similar occurrence took place at the New-York hospital, some months since, to two young gentlemen engaged in sewing up the abdomen of a patient who had been examined by the physicians and surgeons of that institution; each of them pricked one of his fingers in the operation; in a few hours afterwards the part pricked in both these instances, became painful and swelled; the inflammation extended itself along the arm, till it reached the axillary gland, on which it produced violent inflammatory action, which was of some considerable continuance. In one of these cases, the lymphatic vessels were so highly inflamed as to become visible in the course of their distributions along the arm.

Rush‡ notices like affections of this set of vessels, in the Philadelphia epidemic, of 1793. "I met with three cases of swellings in the inguinal, two in the parotid, and one in the cervical glands." And the epidemic of our own city, in 1795, afforded some few instances of this sort.

The appearance of buboes in the plague, as remarked by every writer

* Sect. 1409.
† Preface to Method Chem. Nomenclature, p. 111.
‡ Page 68.

writer on that subject, may be taken as the operation of these causes on these parts of our constitution.

Nor are the brute animals exempted from the influence of these stimulating powers when applied to their œconomy. For *cats* and *dogs* have been known to suffer by the appearance of buboes, &c. as related by the writers on plague, and other like maladies.

Sect. V.—*A new Theory of Fever.*

FROM the operation of these combinations the learned Professor of Chemistry has deduced a new doctrine of fever. Judging from their effects, pestilential fluids appear to be always, even in their weakest form, somewhat of stimulants. In many instances, they are most violently so. Though their operation is modified in a very peculiar manner, when in a dilute form they impede respiration, or nauseate the stomach, as they then bring on a diminution of action and energy, amounting in the cold stage to a state of direct debility: when applied in great quantity and force, they kill instantly; when in less quantity, they produce an anomalous disease, of the form of which Chisholm's cases present us with instances; when in a weaker state, a common contagious catarrhal affection may be the consequence; when inhaled in a form yet more diluted, a remitting or intermitting fever may be the disease induced, of the form of quotidian, tertian, or quartan, or any of their varieties; or the remitting may be called jail, hospital, ship, camp, army, yellow, putrid, or bilious, malignant, pestilential, miliary, petechial, ardent, slow, continued, continual, dysenteric, contagious or infections, according to the circumstances that may occur in the progress of the disease.

" The main difficulty left is to account for the cold fit of a regular tertian. This stage of fever, I believe to depend upon impeded respiration; and the impeded respiration to depend upon the vitiated quality of the air, taken into the lungs; or in some slighter cases, where the stomach is originally thrown into a disordered state, the lungs, by association with that organ, are thrown into disorder too, and for a time perform their functions but imperfectly.

" Thus I presume it is, that the impeded state of respiration is attended with a smaller evolution of heat and oxygene in the lungs; and consequently with more or less diminution in the circulation of the blood, and a proportionable degree of chilliness and coldness throughout the body. The duration and degree of the cold fit will correspond to the continuance and power of the causes disturbing the pulmonic organs, either by acting upon them directly or indirectly, through the intermedium of the stomach.

" From

" From the small quantity of heat and oxygene communicated to the blood in the lungs, and the consequent slow and feeble circulation of the blood, can the shrinking, paleness, tremors, coldness, debility, &c. &c. be sufficiently explained, as the constitution is now deprived of its two chief stimulants.

" But why does not the continued operation of the vitiated air upon the lungs, or the associated condition of the lungs with the stomach, go on in an increasing series even unto death? The power of our constitutions to become familiarized to the action of noxious causes, is evinced by the innocent operation of poisonous substances, which, by frequent repetition, grow gradually habitual, and by custom, lose their primary operation. This disposition to become familiarized to vitiated airs, is apparent in the inhabitants of Africa, who are so seasoned to the air and climate they live in, that it excites no disturbance at all in their constitutions, while strangers fall victims in the greatest abundance. Now, common intermitting paroxysms, are instances of temporary seasonings, which the constitution experiences, of a kind quite analogous to what is perpetual with the Guinea negroes.

" The cold fit sometimes does terminate in death; and this happens when the constitution cannot acquire the habit of enduring the noxious cause with impunity. In the generality of cases, however, the stimulus of the infectious gas loses its power to operate before the constitution is debilitated to death; and as soon as it becomes, for this time, so much accustomed to the vitiated air, as no longer to be disturbed by its presence, the cold fit ends. The length and violence of the cold fit will thus be *cæteris paribus*, in a compound ratio of the impediment given to the respiration by the infectious gas, and the facility with which the constitution accommodates itself to its action; if three persons then inhabit one house, it is possible that one may become so quickly accustomed to the air, as to have no distemper; a second may have a moderate disease of but a few fits; while the third, possessed of a constitution not easily moulded to a new habit, may be incommoded by a violent and obstinate malady.

" In every paroxysm of an intermittent, the infection thus wears itself out; but this is only a temporary reconcilement of the body to its action; when, after a repetition of fits, the disorder becomes milder and milder, and after a while wholly ceases. This is a case of lasting reconcilement; and in this way may a large portion of small intermittents cure themselves, while the credit is given to the bark! This power of habit daily does wonders, and labours more effectually for the good of the sick than bark, opium and antimony put together.

" The

" The attack of these causes being thus for a time overcome, respiration grows free, full, and frequent; because there is now a greater appetency in the constitution for heat; more vital air is decompounded in the lungs, and more stimulus is applied by means of the increased heat and oxygene now in the blood, to the heart and arteries; these stimuli operate more powerfully on account of the accumulated excitability of the body; and a degree of excitement is thence induced which sometimes ends in death, sometimes causes delirium, and in almost every case, exceeds the healthy temperature.

" The duration and violence of the hot stage, will be *cæteris paribus*, in a compound ratio of the excitability accumulated in the cold stage, and the heat and oxygene evolved in the hot one. When the excitability is exhausted by the operation of the stimuli, the violence of action will cease, and the body grow cool.

" The doctrine of intermitting fever then, is briefly this: the vitiated atmospheric fluid, by interfering with the pulmonic action, brings on the cold stage, and would continue the same until its termination in death, did not the constitution, in the meantime, acquire such a habit as to gain a temporary insensibility to its action. This habit being induced, the cold stage abates by reason of the state of direct debility, into which the body had been brought; anxiety continues, and by the quickening of respiration, heat and oxygene are set loose in the lungs, and becoming incorporated with the blood, now warm, and stimulate every part with more than usual power, and occasion the phænomena of the hot stage, which terminates as soon as the accumulated excitability of the system is exhausted.—The sweating stage follows of course, as in other cases of the subsidence of violent action: for after a time, the exhausted excitability of the animal system, allows excessive action to go on no longer, the respiration grows more moderate and easy; the heart beats with less frequency and force; the arterial contractions are also more slow and health-like; and, as the arterial contractions relax, the hydrogene and oxygene of the blood now run together in the extreme vessels of the skin, and form the moisture which bedews the surface, and this afterwards flying off by evaporation, cools by degrees the whole body down to its ordinary temperature: and, as the arterial extremities of the rest of the body become dilated by the subsidence of excitement, the other secretions, which had been generally suspended during the fit, now return as before: after this, the constitution, so far accustomed to the breathing such an atmosphere, regains its former vigour and functions, as far as the exercise induced and functions injured during the several stages will allow.

" The

"The interval between one fit and the succeeding one, will be proportionate to the duration of the habit of resistance acquired. Some persons thus experience but one fit, and all is over; for, under the same circumstances they are never invaded by a second. Others suffer two fits, or a succession of fits, and, after a while become so accustomed to the stimulus, that, if always applied in the same degree of strength, its effect is no longer felt upon the body; in other instances again, so hard is it for the constitution to be moulded into a settled habit of opposition, that after enduring a great number of invasions, it becomes at length enervated and worn down, so much as finally to die exhausted.

"The species of fever, whether quotidian, tertian, &c. will depend upon the readiness or quickness wherewith the offending cause gains a new ascendency over the body, or breaks the habit. And to the mobility of the body, or ease with which the habit is broken, is to be ascribed, as well the frequency of the returns, as the duration and severity of the paroxysms.

"The anomalous cases of fever, which have puzzled physicians to explain, and nosologists to arrange, are thus very naturally accounted for; since, according to the variation of the cause, as the noxious atmosphere may thicken or disperse, will be the variety in the effect produced; and, as there may be infinite gradations of the deliterious cause, there may be endless varieties in the morbid effect.

"And to this principle of the human constitution, I believe may be referred all the febrile ailments from the most trifling intermittent to the more serious remittent, and the solemn form of continued fever.

"Hence further may it be understood how a succession of fits, long continued, may dispose the constitution to a repetition of fits, even when the morbid cause is away: for though there may be a habit of insensibility produced to the vitiated airs, yet a habit may, in the meanwhile, be established in the bodily motions of falling periodically into regular trains of action, even when the original cause is withheld. Here then will be produced a habit of having paroxysms depending on the particular inward state of the moving fibres, after the manner of temporary seasonings; while, at the same time, there is a habit formed of resisting the active causes (vitiated air) altogether, or of obtaining a permanent seasoning as to them.

"The cold stage of a paroxysm is a state of direct debility, induced by the vitiated air breathed operating to subduct heat and oxygene from the body; and its termination is by the stimulus of the vitiated air being for that time worn out. The hot stage, which begins

begins as soon as the temporary seasoning is induced, is a state of excitement brought on by the heat and oxygene now operating upon the accumulated excitability with additional force.—The sweating stage is formed after the subsidence of the excessive action of the body, and the consequent enlargement of the diameters of the vessels, whereby sweat is formed by the combination of hydrogene and oxygene, and the other secretions proceed again, as usual, in the several glands.

"The length of interval between the paroxysms depends upon the strength of habit acquired.

"The frequency of their occurrence will be proportioned to the facility with which a temporary habit is broken or gives way.

"The cold stage is the most dangerous, and persons dying in it die of the direct debility induced by the vitiated atmosphere they respire.

"The hot stage is less dangerous, and persons who die in it expire in a state of indirect debility. But, according to circumstances, death may happen in both the cold and hot stages.

"The sweating stage is a mere consequence of the cooling of the body, after the preceding heat and excitement of it."

The simplicity of this theory, when put in competition with the complex doctrines of preceding physicians, at the same time that it causes our astonishment at their divergency from the fundamental laws of nature, flatters our judgment with its own approximation; and as it does not take the mind into the fancied reasonings of speculative hypothesis, neither does it deprive it of satisfactory reflections on facts and nature. The powers of habit and custom, though hitherto acquiesced in, in almost every action of life, have not as yet been sufficiently considered in their relation to febrile phænomena. Hence have the most useless remedies been brought into estimation, and acquired a reputation to which they had no just pretensions; the disorder abated as soon as the constitution had become habituated to the new stimulus, and not because specifics were administered. Hence, though blood be drawn off to the amount of upwards of an hundred ounces, or calomel administered to the extent of sixteen hundred grains in the course of one febrile indisposition, the patient may get well in spite of both, as soon as the habit of resistance is formed.

APPENDIX.

APPENDIX.

[NOTE A.]

On the Manner in which the Materials of Dwelling-Houses are affected by Septous Fumes and Combinations. In a Letter from Dr. SAMUEL L. MITCHILL, *to Dr.* EDWARD MILLER, *of Dover, Delaware.*

SIR,

SOME paragraphs and essays which have for some time past appeared in our newspapers, and a number of queries proposed to me by my private correspondents, concerning the production of infectious air in houses, and its concealment in sundry substances, of which the habitations of men are constructed, have determined me to collect such facts as occurred to me on that subject, and make them the matter of a letter to you. Your very obliging and friendly favour, dated at Dover, the 15th of November, merited a more speedy answer, but I am confident you will pardon my tardiness, when you consider that I have of late, like yourself, been engaged in an inquiry, tending to alleviate some of the inconveniencies which result from our mode of living, especially in large and populous cities.

And here I shall take it for granted, as proved in my treatise on contagion and elsewhere, that the air of houses, tenements, and dwelling places, is vitiated occasionally in a very alarming and deleterious degree by the gaseous oxyd, or volatilized acid produced by animal and vegetable decomposition, which have been denomited the *nitrous*. I have for some time been of opinion, that an examination of the facts upon this point is an important desideratum, both in philosophy and house-keeping.

The materials of houses may be classed under three general heads, as they consist, 1st. of Earth; 2d. of Wood; and 3d. of Paper.

1. The *earthy* materials of human habitations, whether of stone, brick, or plaster, may be considered as consisting in the main, of flint, clay and lime, the greater part of them being capable of resolution by analysis, into one or more of these elements. The

flinty

flinty parts of a building, comprehending the stone and the sand mingled with the mortar, are from their nature not capable of uniting, in ordinary temperatures, with any of the common acids in any of their forms, and therefore remain pretty much in their original state as long as the house stands.

But the case is far different with the lime and clay. The operation of burning lime-stone to render it fit for the purposes of masonry, is principally to deprive it of the carbonic acid (fixed air) with which, in its natural and crude state, it had been united. And this kind of air is so abundant in the atmosphere, that as soon as the calcined material, now converted to quicklime, is taken from the kiln, it begins to recombine in a slow and gradual progress with the fixed air, which had been expelled by the fire, and returns to the state of mild lime. In this condition, of greater or less combination with its original acid, it is worked up into mortar, and employed as a cement. And after its application to the purposes of brickwork or walls, lime may be considered as attracting its lost fixed air as fast as circumstances will permit; and this process may go on until a greater part of the lime is saturated. In this situation the walls may remain an indefinite duration of time, until the fixed air is expelled by some substance having a stonger attraction for lime than itself.

The habitations of men are known to afford such a substance, and the septous (nitrous) exhalations extricated in rooms, are found to displace the carbonic acid by virtue of a stronger attraction for lime, and attach themselves to that calcarious basis in the form of a nitrate of lime. There are several substances, such as the acids of sugar and of sulphur, which have a stronger affinity to lime than the nitrous acid has; but these rarely occur to disturb the common process. The plaster then by degrees becomes charged with the acid of salt-petre, attracted from the air of the chamber. This operation may go on, in an old and foul house, until the whole of the lime is saturated with the septous vapour, and can take up no more.

These noxious exhalations being now no longer attracted by the lime, must either circulate through the house, or combine with some substance capable of fixing them. The *clay* of the walls is a basis with which they are disposed in the next place to join themselves. The nitrous acid, in the absence of the sulphuric (vitrolic) has the strongest attraction for clay, of any; and the combination of these two substances may thus go on as long as any particles of clay retain a capacity to attract the acid.

In the progression of things, the clay and the lime become loaded with as much septous acid as they can possibly absorb; and after

after this the pestilential gas, finding no other material to form a chemical union with, will be accumulated and diffused through the room or house, penetrating the interstices of bibulous and porous substances.

A fashion prevails in our newly constructed and elegant houses, of making the walls of *gypsum*, or plaster of Paris. This substance is a compound of lime with the sulphuric (vitriolic) acid. The sulphuric acid, as was mentioned before, possessing a stronger attraction for its calcarious basis than the nitrous possesses, cannot be altered or dislodged by the septic vapours circulating in the rooms. The lime of such walls will therefore remain for an indeterminate time, in their original condition; and the foul gases will immediately, and with greater readiness than in the case of lime-walls, vitiate the air of the house.

Walls made of lime and clay, may be thus viewed as preventives of infection, while those constructed of gypsum have no such salutary operation. The reason of the wholesomeness of the former is, they are constantly taking the matter of contagion out of circulation, and fixing it it such a manner as to render it quite harmless. As long as the walls preserve their attractive powers they render the air of a room fit for respiration in proportion to the quantity of gas they imbibe; when they cease to attract any more of it, the air becomes less fit for animal life, by the surplus of gas unabsorbed.

To make houses healthy then, the walls of such as have stood a long time, and have become highly nitrous, ought to be broken down and a fresh plastering of lime applied; or if this could not be conveniently done, a white-washing, which is only a thinner coat of plaster, should be frequently performed. The septic fumes will then have something to attach themselves to, and be taken rapidly out of circulation. An easy experiment will determine whether the walls of houses suspected of being nitrous, are really so or not. Take a quantity of the old plaster, pound it and steep it in water, and add some potash to the mixture; if the nitrous acid is there, it will quit its connection with the plaster and join the potash to form nitre.

The existence of nitrous acid in old grave-yards, at the same time it points out the almost sinful impropriety of burying the dead near the habitations of the living, directs us also to a remedy of the evil. If the coffin containing the corpse was filled up with lime or potash, the danger of communicating infection at funerals, and the unwholesomeness of burying-grounds, would in a great degree be obviated. In like manner, the addition of the same substances to the numerous privies of crowded settlements, would have a powerful effect in preventing the ascent of deadly vapours

into

into the air, or their penetration to cisterns and wells of water through the earth.

2. As to the timber which enters into the fabric of our houses, whether it be oak, pine, cedar, mohogany, or of any other kind of wood, I know of no decisive facts evincing a *chemical union* between it and putrid vapours. But as all these materials are considerably porous, there can be no doubt of their receiving into their interstitial spaces a portion of the vapours which occupy the rooms. The quantity imbibed will probably be in proportion to their sponginess or laxity of texture; and in this ratio may the different kinds of wood be imagined capable of penetration by foul steams. In most cases of this kind however, I apprehend the contagious matter is in a very separable state, and ready on slight occasions to manifest itself in its proper and distinct form. From the readiness with which wood is penetrated by water, and from the known property of water to act as the vehicle to contagion, there can be no doubt of its entering pretty deeply into the timber of our dwellings, especially those that are not covered with a coat of paint. The covering the inside work of houses with paint, by occupying the pores of the wood, will exclude the entrance of pestilential vapours, and thus far tend to their purity and wholesomeness. It is said that ships have sometimes become so contagious as to infect every successive crew with which they were manned. This must in all probability have arisen from the noxious matter inherent in their timber. And I suspect the same thing frequently happens in many ordinary houses inhabited by a succession of dirty tenants. Wood however would seem to admit of easy cleaning. Clear water will go a good way towards purifying it. But the infectious matter may be, without doubt, extracted from it by careful washing with a solution of *potash* in water (common lye) or by white-washing with *lime*, or even *with clay*.

3. The paper with which houses are adorned, is a material of such open and spongy structure, that most fluids can easily enter into it. Oil and grease very readily insinuate themselves into paper. It very readily imbibes water, and in very considerable quantity too. Smoke tarnishes it very quick. And the colours of ancient records and books are not wholly to be ascribed to the decay of the paper and ink, but in part to the impregnation of the pages with foreign vapours. From these facts, shewing the readiness and avidity with which paper combines with most liquid and aeriform bodies, there can be little doubt of its possessing a capacity, like wood and other porous substances, to receive contagion among its filaments. But as paper possesses little or no *chemical* attraction for pestilential vapours, they will not change their

their nature, or lay aside their peculiar qualities, by entering into the interstices of the paper, but remain in a very loose sort of connection, and exceedingly prone to detach themselves and sally forth into action. This idea is countenanced by the facts told concerning the catching of the plague by handling letters, which in these cases were very probably damp, and by aid of moisture, concentered more of the poison within them. Yet when I reflect that the bibulous or absorbent quality of paper is the very cause why it is fit for receiving the stains, marks, spots, and colours which we impress upon it by writing and painting, it seems as if it might attach and retain a considerable share of venom, even in its dry state. Upon the whole, paper may be estimated as possessing all the inconvenience of wood, and in a much more considerable degree. Nor will this conclusion be invalidated by the consideration that the paper is covered with paint, for this layer of pigment is not mixed in *oil*, as in the paints applied to wood, but it is generally *water colour*, and therefore greatly more open and penetrable. Notwithstanding these objections to paper, it may safely be employed in drawing rooms and parlours without any sensible inconvenience or risque, but bed rooms and nurseries, especially in crowded families and infectious situations, had better be furnished with lime walls, and coated over with a calcarious white-wash from time to time: when families are small and situations neat and healthy, it is wholly immaterial in which way the chambers are finished; but where the contrary is the case, it must be remembered that frequently changing the paper, will scarcely prove a remedy of the inconvenience. On the subject of paper, an interesting consideration is, what danger there may be in receiving letters through the medium of the post-office, from infected places. That paper may receive and impart infection, I hold to be a settled fact: and I hold it to be no less a fact, that as there is no chemical connection between the one substance and the other, *heat alone will disengage them*. If a letter, therefore, should be received under circumstances leading to a suspicion of contagion, it should be held as near the fire, with a tongs, as possible without burning it, and continued till it be thoroughly heated. The gas will be rarified and volatilized by the heat, and will chiefly, if not entirely quit its lurking place in the paper, which may be unfolded and read with safety.

With great regard and attachment,
I remain, very unfeignedly, yours,
SAMUEL L. MITCHILL.

New-York. Jan. 20, 1796.
Dr. EDWARD MILLER.

[NOTE B.]

[NOTE B.]

Copy of a Letter from FLEMING MARTIN, *Esq. chief Engineer at Bengal, dated 1st of October*, 1765.

IN regard to the intense and uncommon heat in this climate— it has been for some time past almost insufferable.

The thermometer was seldom under 98, and the quicksilver rose at certain times of the day to 104 degrees by the best adjusted instrument; nay, I have been assured by some gentlemen, that in the camp, 500 miles distance, the thermometer often stood at 120; but such a difference, I imagine, was occasioned by the badness of the instrument.

However, it is certain, that nothing could exceed the intense heat we felt, day and night, during the month of June. May and July were little inferior at times, but afforded some intermission, otherwise very great mortality must have attended this settlement; though we were not without instances of fatal effects in the month of June, when some few individuals, in sound health, were suddenly seized, and died in the space of four hours after; but considering the malignity of the climate, we have not lost many, and I believe the generality of the people are not so intemperate as some years past they used to be; though, from what I have seen, the best constitution, in the most moderate persons, is a poor match against a fever, or other disorders, in this country.

I have been as free from sickness as any other person in the settlement; but I cannot say that I have enjoyed myself in that degree as to be an exception; for no man here is without complaints, and life and death are so suddenly exchanged, that medicines have not time very frequently to operate before the latter prevails. This is generally the case in malignant fevers, which are here termed *pucker fevers*, meaning (in the native language) strong fevers.

The rains have set in since the fourth of June. We call this the unhealthy season, on account of the *salt-petre impregnated in the earth, which is exhaled by the sun*, when the rain admits of intervals. Great sickness is caused thereby, especially when the rain subsides, which generally happens about the middle of October. The air becomes afterwards rather more temperate, and till April permits of exercise to recover the human frame that is relaxed and worn out by the preceding seasons; for in the hot periods every relief is denied except rising in the morning, and being on horse-back by day-break, in order to enjoy an hour or little more before the sun is elevated: it becomes too powerful by six o'clock to withstand its influence; nor can any kind of exercise be attempted that day again till the sun retires, so that the rest of the

the twenty-four hours is past under the most severe trials of heat. In such a season, it is impossible to sleep under the suffocating heat that renders respiration extremely difficult: hence people get out into the virando's and elsewhere for breath, when the dews prove cooling, but generally mortal to such as venture to sleep in that air. In short, this climate soon exhausts a person's health and strength, though ever so firm in constitution, as is visible in every countenance after being here twelve months. I have lately been informed by an officer of distinction, who was formerly an engineer at this place, that being sent out to survey a salt lake in the month of September, he found the sulphurous vapours so stagnated and gross, that he was obliged to get up into the tallest trees he could find to enjoy the benefit of respiration every now and then; he added, that he constantly had recourse to smoaking tobacco, (except during the hours of sleep) to which, and to swallowing large quantities of raw brandy (though naturally averse to strong liquors) he attributed his safety. However, on his return, he was seized with an inveterate fever of the putrid kind, which he miraculously survived; though others, who attended him on the survey, and lived many years in the climate, were carried off at the same time by the like fever.

[NOTE C.]

On the Miscibility of Contagious Air with Water. In a Letter from Dr. SAMUEL L. MITCHILL, to JOHN STEVENS, Esq.—Read before the Agricultural Society of the State of New-York, February 16, 1796.

DEAR SIR,

AFTER I had satisfied myself of the composition of that particular kind of air, which is produced from substances putrefying in places where *heat* and *moisture* concur to form new combinations, my next object was to detect its particular qualities, and relations to other bodies. And I have become persuaded that in all its forms, this oxyd or acid of septon, is *readily and entirely miscible with water.* By water I do not mean the pure distilled fluid merely, but also the water of rain, snow, ponds, and marshes; and occasionally of the ocean, as well as the fluids which, though called by other names, do still consist chiefly of water, as vinegar, wine, spirits, &c.

Some of the interesting considerations which arise from this chemical attraction between watry fluids and infectious air, I shall endeavour to state to you in the order in which they rise to my recollection,

recollection, in answer to your polite and instructive letter, dated at Hobocken, September 25, 1795. In dependence upon this single principle, a multitude of facts press upon the mind, and all appear equally easy of solution.

1. It has been experienced in the neighbourhood of unhealthy landings, in the East and West-Indies, that ships lying a small distance off at anchor in the harbours, or at sea, have in no degree suffered by the noxious quality of the air, which was very injurious to the settlement on shore. But there, as well as on the coast of Africa, a boat's crew sent on shore, and staying the night, are very commonly seized with sickness, which very commonly destroys a number of their lives. The reason of which seems to be, that the contagious air (septous oxyd or acid) produced from putrefaction on the low and muddy banks of rivers, &c. is most concentrated at the place of its origin. Here, sailors tarrying long ashore, are obliged, in uncleared countries, to hut themselves, and sleep near the ground. From the greater accumulation of contagious air near the surface of the earth, they must be more particularly exposed to its action while lying down than when setting up or walking about. By remaining for six or eight hours in this manner immersed in such an atmosphere, and during the time of sleep too, one may readily understand how they either sicken on shore, or soon become unwell after getting on board. They suffer by its action in four ways: 1. By *landing* where the atmosphere is vitiated. 2. By *lying near the ground* where the noxious gas is most condensed. 3. By remaining *so long a time* within the sphere of its operation. 4. By *going to sleep*, that it may damage the body in that unguarded state. In the vicinity of all this destructive air, the ships escape for two reasons: 1. Because, if the poisonous air should reach them, it would be in a rare and dilute form. 2. Because, in *passing over the water, so much of it would be absorbed*, that a very thin stratum would remain on the surface, not many feet, or perhaps inches in height; so that little or no inconvenience could result from it. Hence the service of cutting wood, getting water, burying the dead, trading excursions, exposure in open boats along coasts, &c. are extremely hazardous in many hot climates.

2. The experiments made at Portsmouth, Cadiz, Sardinia, Pensacola, and in Guinea, have proved that ships anchored off a little distance from places where mortal sickness rages on shore, afford a most convenient retreat for those who are well, and allow an excellent chance of recovery to such as are sick. The security of such places in unhealthy seasons, has led, very naturally, to the idea of recommending *floating factories* to the merchants who have trading

trading establishments on the sickly coasts of Africa, as at Cape-Coast-Castle, the mouths of the rivers Senegal, Gambia, &c. Mr. Dodge, a valetudinarean, who built *an ark*, as he called it, for himself and family, to avoid the influence of the sickly season on shore, has shewn, by experiment, how well a private gentleman may be accommodated and enjoy his friends in a *floating mansion*, consisting of a sleeping-room, dining-room, apartment for servants, kitchen, coal-room, wine-cellar, &c. The principle upon which such places of abode are pronounced healthy is, that as the contagious fluid is produced ashore, or from mud in the neighbourhood, a ship or *floating habitation* cannot commonly be affected by it, on *account of its attraction for water*. The ideas mentioned in your letter are therefore rational and just, "That in seasons of infection, temporary stages be framed of dock-logs, and anchored off the city, at a convenient distance from the wharves. On these suitable apartments might be erected for the reception and accommodation of the sick. And, in order to prevent, as much as possible the accumulation of contagion, let the stages be multiplied as much as they conveniently can, and placed at proper distances from each other. The most important benefits would probably result from an arrangement of this nature. By having a number of them, situated as they would be, it would be very practicable to keep them free from contagion, as they might occasionally be suffered to remain unoccupied long enough to cleanse them properly. The sick would be moved with the utmost ease and convenience, without the aggravation of their cnmplaints by the jolting of a wheel-carriage. The dead might instantly be disposed of, and the danger of spreading contagion, and panic terror, by conveying dead bodies through the streets in a hearse, would be prevented. As the removal of a sick person to an hospital of this kind would probably prove extremely salutary to himself, there would not be that reluctance either on the part of the sick person or his friends, to this necessary measure; and thus the efforts of the police to separate the sick from the well, so far from meeting with opposition, would be aided and assisted by every body. The opulent, no doubt, as a necessary precaution, would provide hospitals of this sort at their own expence. Thus situated, every comfort, aid, and convenience might be afforded the sick, and their friends might attend them with scarcely any apprehension of danger. These *aquatic lazarettos* might also be employed in another way, probably to great advantage. The air of the city, during the hot months, is extremely injurious to young children. Vast numbers are carried off every year, during the summer season, by disorders of the bowels. It is truly astonishing what an immediate effect, a removal from the foul air of the city has on infants labouring under these complaints."

3. The remarks of almost all observers agree in this point, that the aerial vapours from stagnant water are seldom mischievous, until evaporation has so far advanced as almost, or quite, to bare some part of the mud or bottom. The reason of this is evident. As long as evaporation goes briskly on, the evaporating surface is kept cool, and putrefaction of course, advances but slowly. The gas extricated below, mingles with the incumbent water, and for some time, little or none escapes to taint the atmosphere. But as the quantity of water shrinks, and the swamp or pool dries up, the attraction of the atmosphere for the water being stronger than that of the water for the contagion; the latter is discharged from its connection, and floats about at large: the mud now grows warmer, putrefaction advances, more infected air rises, and there is less and less water to arrest its ascent. And thus the process is carried on, until the want of sufficient moisture, to promote the chemical action of bodies, puts a stop to the work, and sends forth all the gas into the atmosphere. This explanation corresponds exactly with the facts relative to its operation upon human bodies, and the consequent production of sickness.

4. Rain falling briskly in showers is found to have a beneficial effect in seasons of contagion. Some of the most infectious distempers we are acquainted with, have happened in times of great drought. The suffering inhabitants of sickly places have experienced sensible relief from showers, and have generally relapsed when dry weather came on. To understand this, it may be proper to consider the infectious fluid as having risen some distance, more or less above the ground, and the drops of rain passing through it as they fall. In a condition so favourable for union, the rain attracts the contagion, and carries it down to the earth. The sick are thereby relieved, and the well less exposed to danger. By and by the water evaporates, and leaves the infectious fluid by itself to rise again and contaminate the air—Mortality recommences. The natives of Africa are so sensible of something unhealthy mixed at times with water, that though they generally bathe once a day, they never do it in the fresh river waters, when they are overflowed by the rains; but prefer spring-water, which flows pure from the interior parts of the earth. The *first* rains which fall in Guinea, are supposed to be the most unhealthy, and as evidence of their being impregnated with something more than common, they have been known to render the leather of shoes mouldy and rotten in forty-eight hours, and to stain clothes more than any other rain. Exposure *to rain*, and *getting wet*

thereby,

thereby, are generally believed, says Hunter, to be productive of fevers in the island of Jamaica. I have heard fishermen remark, that exposure and wetting in a shower of rain, was more likely to make them sick, than the regular attendance of their seins, up to the waist or arm-pits in salt water, day after day. There is one case, however, in which a fall of rain may be productive of sickness. I shall give an example, as it occurs in Africa. During the drought, the wide rivers of that parched continent, are shrunk to narrow streams, and retiring to their channels, leave bare a large part of the surface they usually overflow. The moisture soon exhales, and leaves behind it a thick and solid crust of dried mud. When the rains fall, they penetrate and soften this parched crust, soak into the ground below, and set the whole into a putrifying state. From the surface, which, during its dry condition, no vapours had risen, now proceed noisome and noxious exhalations, which produce the most violent sickness.

5. Clothes wetted with water, and hung to windward, are the only known correctors of the violence of the *Harmattan* and *Samiel* winds. They may operate in two ways: 1. By mitigating the excessive heat of these blasts by evaporation; 2. By attaching and fixing the pestiferous matter they contain, or turning it aside by reason of its inability to pass through a coarse wetted cloth. A great variety of charms and preventives are used against infection. None of the smelling-bottles, bags, &c. for the nostrils, seem to be of any service. If any thing of this kind applied to the nose and mouth, can do any good in pestilential states of the air, it is *a cloth or sponge wetted with pure water*, or some watery fluid, which will allow the respirable air to pass through or along it, and imbibe infection among the water inherent in the threads or pores. But such a sponge or cloth should be often washed or changed, and never suffered to dry; otherwise the collected contagion set loose on the drying of the cloth, will be suddenly breathed in greater quantity than if no such thing had been used. It is, therefore, very questionable, whether *even this application* would, upon the whole, be of any real utility.

6. Health has long ago been considered as greatly influenced by the quality of water. The saline, earthy, metallic, and to a considerable extent, the *aerial* qualities possessed by this fluid, have been a great way investigated. From the readiness of contagious gas to combine with water, there cannot remain a doubt of its being conveyed into the stomach in large quantities, together with our drink. Though this remark will only apply to the water swallowed cold; for all that which is taken warm in teas, soups, &c. has been deprived of the pestiferous air, by the operation

tion of a boiling heat. It will be observed that I am now speaking of fountain, spring, and well water, as employed for domestic uses; and not of rain water, which certainly contains it. Fresh drawn water may be generally considered clear of it; but water that has stood long in an infected house may be considered very suspicious at best. During the extensive and mortal distemper which prevailed in the vicinity of the Salt-Lake, the Seneka-Lake, and on Genesee, in the autumn of 1795, there was a remarkable connection between sickly settlements and bad water. Where the water was good, few people were unwell. From the description of the uniform shore of the Chesapeak-Bay, where springs of pure water are scarce, and where stagnant pools are filled with frog's spawns, musquitoes, &c. it would seem that the foulness of the water entered largely into the remitting fevers of that region. And I think the same observation applies to other tracts of country, similarly circumstanced all the world over.

7. Spittle and the fluids of the mouth, consist chiefly of water, and therefore may be imagined to possess an attraction for septic and contagious gases. The matter of infection thus getting into the mouth, will almost unavoidably find its way into the stomach with the spittle, in common swallowing, or during the deglutition of the food, or be conveyed thither with the drink. By this inlet many physicians have supposed contagion to be received into the body. The opinion is probably just. A gentleman of my acquaintance received the fumes of a dysenteric evacuation; some uneasiness at the stomach came on, with qualmishness, which in a few hours ended in a looseness. A similar indisposition was brought on in another person by receiving the effluvia at a funeral, from a corrupting corpse. In both these instances, it looks as if the infectious gas was swallowed with the spittle, and thus stimulated the intestines. Nor do the alledged experiments of swallowing without detriment the poisons of small-pox, the viper, &c. in very small quantities, prove that *all other* combinations and modifications of contagion must be harmless. These experiments, though true, are, however, no further true than respects these species of venom; and I own I should like to see them repeated in larger doses. Be this matter, however, as it may, one wholesome inference results from the facts before us, that it is prudent to cleanse the mouth of all spittle, to hawk the phlegm from the throat, and throw them out frequently, and to avoid swallowing those fluids as much as possible in infected places From the various affections of the mouth and throat, in many fevers, there can be little doubt of the contagion's having a deeper and more serious operation, than upon the mere secreted fluids.

fluids. Do not aphthæ, sore-throats, erythematic affections of the pharynx and gullet, arise from the local operation of this poison? This manner of accounting for these symptoms, is, in my judgment, more accordant to fact and reason, than any thing I have yet met with; for, as to the received notions of *eruptions* coming from the blood and inward parts, and *breaking out* upon the surface of the body, I think them a very lame part of our pathology.

8. A long continuance of fogs, with damp and drizzling weather, has been found unfavourable in sickly seasons. The gradual and slow precipitation of water from the air, keeps the atmosphere so moist, that the small drops of water floating in it enter the lungs with every inspiration, and carry a portion of contagion along with them as far into that organ as they penetrate. A damp air may thus be said to give strength to contagion, though a dry wind may spread the mischief.

9. From the same principle may be deduced an easy method of purifying infected goods, apparel, furniture, &c. the most efficacious and easy method of cleansing which, is to wash them in fair water. For this purpose the water should be *cold*, that is, of a common summer temperature, at which range of heat the contagion readily mingles with it. Linen and other things intended to be washed, should never be put immediately into *warm* or *hot* water, as the steam arising from thence, is found, by experience, to be offensive and dangerous. After steeping and soaking awhile in cool water, the goods may be washed out, and *afterwards* put into boiling water, if judged necessary, with safety. The application of this practice to the cases of persons actually sick, to the purification of bedding, houses, &c. where sick persons have laid or died, and to preventing the introduction of contagion from abroad, is as plain as it is extensive; and this principle is corroborated by the fact mentioned in your letter, "That Europeans residing in Constantinople, Grand Cario, &c. have found, from long experience, that by passing every thing they receive, such as food, &c. through water, it is effectually deprived of its power of communicating contagion; or in other words, the pestilential gas, adhering to its surface, is retained by the water."

10. In many cases of sickness, especially where patients are clad in woollen, and lay on feather beds, covered by blankets, an atmosphere of contagious vapours may be imagined to surround and closely invest them. The skin, especially of those not accustomed to purify it frequently by bathing, is beset by the same fluid, whose particles seem to inhere in its pores and duplicatures. The long continued immersion of the sick in shirts and sheets, ren-

dered

dered damp by perspiration, is thus not only a most unclean but infectious practice. The relief afforded to persons labouring under indisposition from infection, by plunging into cool water, or what I prefer, wiping the whole body with a wetted cloth, again and again, arises not so much from the abstraction of heat as by removing the irritating and pestilent matter adhering to the surface. The purification of the whole body, by clean water, appears to me to be as necessary, in infectious ailments, as cleansing the stomach and bowels by internal remedies; and, I had like to have said, in some instances quite as beneficial. If the observance of personal cleanliness in this respect was more frequent, and chamber-baths, as among the French, more generally used as a part of bed-room furniture, contagious distempers, I am persuaded, would be far less frequent. The notion of its checking perspiration being founded on a false estimate of the animal œconomy, is too idle to listen to or to refute: and what adds further to the truth of the principle stated is, that washing the bitten part in cold water, has been recommended as very efficacious in preventing the consequences of the bite of rabid animals; and the latest West-India treatment of yellow fever confirms the utility of the boldest use of cold water, by dashing it over the head and shoulders with a bucket; or if the infected person was unable to set up, by wrapping him from head to foot in a blanket dipped in cool salt water.

Thus, Sir, we learn that if the ingredients of atmosphoric air, (septous and oxygenous gas) the commonest things in nature, do sometimes get into chemical combination, and produce a pestilential or non-respirable fluid, there is also another thing, one of nature's most plenteous productions, which seems in its pure state to be a sovereign preventive of a large proportion of their mischief.

With deep esteem and respect,
Believe me to be,
Very sincerely yours,
SAMUEL L. MITCHILL.

Plandome, *Feb.* 6, 1795.

JOHN STEVENS, Esq; Hobocken.

[NOTE D.]

[NOTE D.]

On the Decomposition of Contagious Air by Vegetation. In a Letter from Dr. SAMUEL L. MITCHILL, *to* ROBERT R. LIVINGSTON, *Esq; Chancellor of the State of New-York, and President of the Agricultural Society.—Read before the Agricultural Society of the State of New-York, February* 23, 1796.

DEAR SIR,

EVER since the idea has been started of epidemic and endemic diseases being caused by liquid or aeriform compounds of the principle of *putridity*, (septon or azote) with the principle of acidity (oxygene) in different proportions, it has appeared to me an inquiry highly worth the making, by what process in nature such vast quantities of noxious fluids are decompounded, or taken out of circulation. In my letter to Dr. Miller, of Dover, dated January 20, 1796, I stated the facts which occurred to me, concerning the absorption of this class of fluids *by the materials of which houses are built*, particularly as respected the *lime* and *clay* of their walls; and in another communication of Feb. 6, to Mr. Stevens, of Hobocken, I pointed out the considerations then present to my mind, relative to their *miscibility with water*. By these modes of combination, some of those deleterious fluids are so damped or repressed, that their hurtful powers are quite suspended, until, by a severing of the connection they had formed with some other substance, as with lime, clay, water, &c. they break out into action again.

Besides these processes, in which the contagious airs are fixed or connected, so as to disappear for a time from circulation, there is yet another operation in nature which is capable of utterly destroying them, and resolving them into their constituent elements. This operation is VEGETATION.

I was induced to this opinion by the following circumstance: The scavengers and cartmen of the city of New-York are in the habit of collecting from the streets large quantities of putrid mud and filth, and of laying it in heaps on vacant lots, near the skirts of the city, to be converted into manure. Near the ship-yards, on the East river, in the neighbourhood of a number of these collections of street manure, the yellow fever was very rife and destructive; and in the opinion of the most observing inhabitants, a considerable share of its violence was owing to their vapours and exhalations in the autumn of 1795. That putrid effluvia should cause sickness, is a very ancient idea; but the acquiring a precise idea of the nature of putrid effluvia, has at once such an intimate connection,

connection, both with agriculture and physic, as to render it a matter of more than ordinary importance.

These heaps of manure appear to me to be *nitrous* soils, and capable, under proper management, of affording much nitrous acid. The effluvias from nitrous soils in Bengal (Martin's Letter in Meigs's New-Haven Gazette, May 22, 1788) and Spain, (Townsend's Travels) are themselves nitrous and very noxious. The nitrous nature of putrid vapours appears from Mr. Thouvenel's experiments, (Memoirs de l'Acadimiæ Royale des Sciences, &c.) and I think from Dr. Priestley's (2d Exp. and Obs. p. 217.)

The collections of putrifying substances in the parts of New-York more especially affected by that distemper, may be considered as so many smaller dung-heaps, and generating in like manner a vast amount of nitrous exhalations. These, when afforded sufficiently concentrated and copious, by the concurrence of sufficient heat and moisture, produce yellow fevers, bilious fevers, &c. and occasion great panic and alarm in society.

Now, these very substances that cause so much mischief and terror in cities, are sought after with great avidity by farmers, who purchase them at a high price, and use them with much advantage to fertilize their fields. The beneficial and salutary effects of this practice in husbandry, makes it look as if nitrous acid and nitrous airs were good manures, and that vegetables had the power of decompounding them. That in short, in the œconomy of plants, there is a process by which the septon and oxygene of these infectious fluids are separated, and, while the former remains in the vegetable, as a part of its nutriment, the surplusage of the latter, after forming gum, mucus, meal, &c. and other vegetable oxyds, flies off through the upper surface of the leaves, in company with heat and light, in the form of vital air.

In order to establish this conclusion, it will be necessary to show, that vegetables contain the principle of putridity. The most ready way of coming at this will be by examining whether they contain ammoniac (volatile alkali.) Ammoniac has the same radical (septon) with nitrous acid, but the two compounds differ from each other in this, that to form ammoniac, septon is combined with hydrogene; whereas in the composition of nitrous acid septon is connected with oxygene. If the volatile alkali then can be produced from plants, it will be a proof that the radical of the nitrous acid (septon) is an ingredient in the vegetable structure.

Pit-coal is generally considered as a substance of vegetable origin. In the treatment of coal, to procure tar from it, by Lord Dundonald's process, a large amount of volatile alkali is obtained. That whole class of vegetable bodies called funguses, or the toad-stool tribe,

tribe, yields on analysis by fire, the volatile alkali. The whole class of the tetradynamia, to which cabbage, mustard, hore-radish, &c. belong, are supposed to contain or to afford a large proportion of it. The glutinous part of meal, flour, or bread-corn, yields it also, by distillation; and indeed most plants, urged by a strong fire, in close vessels, or when converted into soot, give out the same alkaline product. The case of soot, which affords a good deal of ammoniac, is very impressive, as it seems to evince the formation of the volatile alkali, by the junction of septon with hydrogene in high temperatures, when at the same time there is incomplete combustion.

There is an experiment of Dr. Priestley's, which shews the analogy between nitrous acid and volatile alkali. He found (Exp. and Observat. vol. ii. p. 43—44) that iron, which had long rusted in nitrous air, gave out a strong smell of volatile alkali. In this experiment, the iron attracts a part of its oxygene from the nitrous air (which always contains a quantity of water) and converts it into azotic gas and azotic oxyd. The iron likewise attracts a portion of the oxygene from the water, and sets loose a quantity of hydrogene gas. Meantime the metal grows rusty, or, in other words, is converted to an oxyd. But while this goes on, or afterwards the component ingredients of volatile alkali are evolved, and thus the azote and hydrogene combine to form ammoniac. An experiment of Mr. Milner's is very much in point. By exposing alkaline air to substances containing oxygene, in a red heat, he produced nitrous gas, (Phil. Trans. vol. lxxix. p. 300.) Here it seems, the hydrogene of the ammoniac quitted its connection, by reason of some more powerful attraction, and afterwards its radical (azote) combined with the principle of acidity to form nitrous air. We see, thus, how the azotic basis, by combining with oxygene, forms a nitrous, or with hydrogene, an alkaline product.

There is thus no difficulty in conceiving whence vegetables get their azote. Nitrous acid, and some other combinations of azote with oxygene constitute a large proportion of the fertilizing part of street manure. The powers of vegetation are capable of dissolving the connection between these two materials and the oxygene evaporates, while the azote attaches itself to the plant, as one of its nourishing materials. It has been said that volatile alkali exists, naturally formed, in some plants, and nitrous acid in others; but these assertions require farther confirmation by experiment. The better opinion seems to be, that they are both formed on, or by the putrefaction and distillation of such vegetable bodies as contain their constituent parts. The nitrous acid, and the analogous products, being commonly produced by common putrefaction, in ordinary summer heats; and the volatile alkali in high ranges of artificial heat, in distillation and incineration.

It seems (3 Priestley's Exp. and Obs. p. 415) that charcoal is, in some cases, capable of affording azotic air; but this only happen when, after expulsion of all the air it originally contained, it has been afterwards suffered to imbibe atmospheric air. On expelling this secondary portion, there is no carbonic acid procured, but all is azotic air. Its oxygene is probably absorbed, and, as Bancroft very ingeniously conjectures, (1 Philosophy of Permanent Colours, &c. p. 48) forms the black oxyd of charcoal.

It may appear like a strange idea, that nitrous acid and its vapours and airs, should act as manures, and assist in fertilizing land. It is, however, a certain fact, that this is the case; and, on the rapid evaporation and absorption of these compounds when strewed over the land, can it be explained why street-manure has so little durability, and leaves so soon the fields on which it had been spread in a state of exhaustion. In like manner it becomes easy of comprehension, why spots too highly manured, as well as heaps of manure themselves, possess the fertilizing ingredient in a state too strong and condensed to support vegetable life at all. In order to operate well, and produce its greatest good effects, this manure should be tempered with common earth, and by no means be spread too thick upon the ground Unless this be attended to, the manure will corrode or poison the plants it is intended to stimulate and nourish.

Street dirt and dung thus contain nitrous acid. Vegetables absorb it, decompound it, and attach its basis, while they exhale its oxygene through their leaves into the atmosphere. This basis, in *common putrefaction*, is ready to combine with oxygene into nitrous acid again; and in *high heats* to connect itself with hydrogene into volatile alkali.

That some extrordinary process of putrefaction went on during the pestilential autumn of 1795, in New-York, was evident, not only from the general luxuriance of most vegetables, but from the production of a second crop of flowers by many plants which commonly bloom but once in a season, particularly pinks, apple-trees, and cherry-trees. Notwithstanding the advantages derived to mankind from the destruction of vitiated air, and the production of vital air, by the œconomy of plants, the police of the city of New-York, a few years ago, caused a large number of the trees growing throughout the streets to be cut down! And in the same spirit of proceeding, the committee for preventing the introduction of infectious diseases, though waited upon in person, in

consequence

consequence of information solicited by them on the subject of contagion, during the pestilence of 1795, deemed the principles of the present communication not worth the hearing! Is it wonderful we are visited by plagues?

With sentiments of esteem and respect,
I am, Sir, your very humble servant,
SAMUEL L. MITCHILL.

New-York, Feb. 20, 1796.

[NOTE E.]

The Power of Oxygene and Azote to colour Animal Substances. Extracted from "Experimental Researches concerning the Philosophy of Permanent Colours, &c." vol. i. *p.* 30—32. *By* EDWARD BANCROFT, *M. D. F. R. S.*

WE are at this time well acquainted with the constituent parts of the acid of nitre: it undeniably consists of what the pneumatic chymists term azote (phlogisticated or nitrous air) rendered acid by its combination with a certain portion of oxygene, or the basis of vital air. When the azote and the oxygene are combined in a certain proportion, the acid or compound is colourless, as we see it in aqua fortis or nitric acid: but if this colourless acid, in a transparent glass vessel, partly filled, be exposed to the rays of the sun, or the light of a fire, an alteration will take place in the proportion of its ingredients; since the light will combine with a part of the oxygene, and cause it to become elastic and fly off, and the azote will consequently predominate in the remainder; which, merely in consequence of this predominance, will assume first a yellow, then an orange, and afterwards a high vivid aurora, and even a red colour, intensely affecting the sight. But if the glass vessel containing the colourless nitric acid, were filled with it, no such change of colour would take place by any degree of exposure to the sun's rays or other light; because, in this case, there would be no sufficient space or room to allow of a separation and escape of the oxygene. When nitrous acid has been made to assume the colours as before mentioned, if the glass vessel containing it be hermetically sealed and kept for some time in the dark, the oxygene by losing its light, will lose its elasticity; and being again re-absorbed by the nitrous acid, the latter will become colourless, as before. Mr. Keir mentions an orange-coloured nitrous acid, which, by long keeping, became green, and afterwards

of

of a deep blue; and Bergman says, that if, to a concentrated red nitrous acid, one fourth part of the quantity or measure of water be added, the colour will be changed to a fine green; or to a blue, by the addition of an equal measure of water, and that double its quantity of water will destroy the colour. Here then we have an example of all the various colours produced by the two species of air which compose our atmosphere (almost wholly) when deprived of their elasticity, and mixed in particular proportions with more or less dilution by water.

In the same manner, colourless nitric acid, when applied to wool, silk, fur, or the skins of animals, their nails, horns, &c. renders them all, not only yellow, but orange, and even aurora-coloured. Mr. Berthollet thinks these changes are produced by a kind of combustion; but I am pursuaded they are the result of a combination of the oxygene with the azote, which he has proved to be a constituent part of all animal substances; they being exactly similar both in their nature and origin, to the changes of colour produced as before mentioned in the nitrous acid. Were these colours the effect of combustion, why are they not likewise produced in the same manner upon linens, cottons, and vegetable substances, which contain either little or no azote, but a great portion of the basis of charcoal, and ought therefore to be more liable to be acted upon in the way of combustion, than animal substances?